ROUTLEDGE LIBRARY EDITIONS: SOVIET SOCIETY

Volume 17

SCIENCE POLICY IN THE SOVIET UNION

SCIENCE POLICY IN THE SOVIET UNION

STEPHEN FORTESCUE

Routledge
Taylor & Francis Group
LONDON AND NEW YORK

First published in 1990 by Routledge

This edition first published in 2025
by Routledge
4 Park Square, Milton Park, Abingdon, Oxon OX14 4RN

and by Routledge
605 Third Avenue, New York, NY 10158

Routledge is an imprint of the Taylor & Francis Group, an informa business

British Library Cataloguing in Publication Data
A catalogue record for this book is available from the British Library

ISBN: 978-1-032-86028-2 (Set)
ISBN: 978-1-032-88976-4 (Volume 17) (hbk)
ISBN: 978-1-032-88978-8 (Volume 17) (pbk)
ISBN: 978-1-003-54062-5 (Volume 17) (ebk)

DOI: 10.4324/9781003540625

Science Policy in the Soviet Union

Stephen Fortescue

London and New York

First published 1990
by Routledge
11 New Fetter Lane, London EC4P 4EE

Simultaneously published in the USA and Canada
by Routledge
a division of Routledge, Chapman and Hall, Inc.
29 West 35th Street, New York 10001

Typeset by LaserScript Ltd, Mitcham, Surrey
Printed and bound in Great Britain by
Biddles Ltd, Guildford and King's Lynn

British Library Cataloguing in Publication Data

Fortescue, Stephen, *1952–*
Science Policy in the Soviet Union.
1. Soviet Union. Science
I. Title
509'.47

ISBN 0-415-02379-3

Library of Congress Cataloging in Publication Data

Fortescue, Stephen.
Science Policy in the Soviet Union / Stephen Fortescue.
p. cm.
Includes bibliographical references.
ISBN 0-415-02379-3
1. Science and state – Soviet Union. 2. Research – Soviet Union – Management. 3. Research – Government policy – Soviet Union.
I. Title.
Q127.S696F677 1990
338.94706--dc20 89-27682
CIP

Contents

Tables

Preface

Some excellent general studies of Soviet science have been published in English in the past. These include the volume by Zaleski, Kozlowski, Wienert, Davies, Berry and Amann, *Science Policy in the USSR*, the collection edited by Thomas and Kruse-Vaucienne, *Soviet Science and Technology: domestic and foreign perspectives*, and Paul Cocks' 1980 study for the US National Science Foundation *Science Policy: USA/USSR*, vol.2, *Science Policy in the Soviet Union*. But these books were all published some time ago and are now difficult to obtain. There seems therefore to be a need for a new general examination of the management of Soviet science.

Before describing what this book sets out to do, it would be as well to mention what it does not set out to do. Firstly, it is not a comprehensive examination of every science-related institution or issue. The aim has been to deal with the most important institutions and issues, but to sacrifice comprehensiveness for the sake of manageable size and readability. Secondly, it is not a handbook for Western scientists looking for specific information on particular Soviet research or institutes. Those looking for the addresses, telephone numbers, research interests, etc. of Soviet institutes (and a lot more very useful information) are referred to the excellent volume in Longman's Guide to World Science and Technology, M.J. Berry (ed.). *Science and Technology in the USSR*.

So what then is the book about? It is intended to be a general treatment of the organisation, planning and conduct of science in the USSR. The primary stress is therefore on the management of science, rather than its content. The aim has been to write for the interested non-specialist, probably a Western scientist or research manager, although I certainly hope that it will be of interest and value to the specialist as well. I recognise that I cannot hope to be entirely successful in conveying to the Westerner what it is like to do research in the USSR – I have neither the intimate knowledge nor the writing skills to make that possible. Nevertheless it is hoped that the reader will gain some sense of how Soviet science is run and what it is like to work in a Soviet research institute.

It has not been my intention to push any particular theme in the book. However, I do see myself writing in a particular context, and that is that there are enough similarities between Soviet and Western scientific endeavour that a Western reader will understand what the book is about and feel some familiarity with it, even in its more detailed and technical parts. But at the same time there are significant differences that need to be drawn out and if possible explained. I hasten to add however that the book itself is in no way comparative. It contains no more than the odd comment on Western science. I therefore leave it to those more knowledgeable of Western science to draw comparative conclusions.

The context therefore is comparability without the comparison. If there is a theme, or refrain, through the book it is that Soviet science faces serious problems. To some extent I have made a conscious effort to get away from a tendency among Western writers on Soviet science and technology to assume that Soviet research is essentially adequate and that explanations for any problems that the USSR might face in its technological modernisation programmes are to be sought closer to the production end of the spectrum. Some will undoubtedly feel that far too much emphasis has been put on the problems, and that the successes of Soviet science and the reasons for them have been neglected. There might be some truth to that, but I nevertheless feel that the overall balance is justified. That does not mean that I deny the existence or importance of the successes. It is basic to the context of comparability that Soviet science occupies a legitimate place in international science. That in itself suggests an impressive performance.

I face the same problem that everyone writing on the contemporary Soviet Union does at the moment. It is extremely difficult to strike a balance between a description of the well-established features of the traditional Soviet system and the sometimes extraordinary changes which have been so evident in recent times. Any writer at present runs the risk that his work rapidly becomes obsolescent, either because Gorbachev succeeds and the old elements of the system become irrelevant, or because he fails and the exciting new developments are suddenly reclassified as 'hare-brained schemes'. I have tried to deal with the most important of both the old and the new, and can only hope that my judgement does not let me down.

Most of the research for the book was done at the Centre for Russian and East European Studies (CREES) of the University of Birmingham, while I held a Research Fellowship funded by the Economic and Social Research Council (Reference No. F00 232070). I am grateful to the ESRC for its support. I am particularly grateful to everyone at and connected with CREES for their cooperation, stimulation and friendship. The book was written at the University of New South Wales, and I gratefully acknowledge the interest and assistance of my colleagues in the Faculty of Arts and School of Political Science.

Stephen Fortescue
Kensington, NSW

Chapter one

Investment and return

Measuring the size of the Soviet research effort is an immensely difficult and complex exercise. The Soviets release only a limited amount of data, and the data they do release is more often than not incomplete, ambiguous and certainly not comparable with Western data. Western analysts have expended an enormous amount of effort on the data, and all I can hope to do is summarise such work.

Funding

The Soviet definition of scientific expenditure excludes some important categories, particularly most research and development (R&D) carried out in-house by enterprises (although that is a far less important category than those familiar with Western conditions might expect), university research funded by the universities (again not an enormous category), and a considerable proportion of the defence R&D budget. It includes other categories which would not normally be included in Western definitions, particularly social science and humanities research.[1] As well as figures on total science expenditure Soviet statistics also include figures for the science budget, i.e. money allocated directly to science in the government budget. In recent times about half of science expenditure comes from the state budget. The rest is financed from the internal funds of institutes and their customers.

Funding of science in the Soviet Union has generally been quite generous. It is an area fraught with statistical difficulties, but it seems probable that since the mid-1960s at least the USSR has committed more of its national income to science than the USA and other Western countries.[2] However, Soviet national income is so small relative to its Western competitors that it has difficulty matching Western funding commitments.[3] One therefore finds evidence of serious shortages of funds, which tend to be reflected above all in basic infrastructural shortcomings. Funding difficulties were particularly evident in the mid-

1970s, when growth in both budgetary allocations and total science expenditure showed marked declines. There has been some recovery since then. In his speech at the 27th Party Congress Gorbachev recognised that there had been some neglect of the 'material base' of science in the recent past. There have been since then quite substantial increases in the projected levels of the science budget and overall science expenditure (Table 1.1). Projections for the Twelfth Five-Year Plan (1986–90) show total science expenditure (excluding capital expenditure) reaching 33 billion rubles by 1990, as against 24.8 billion in 1985. This is said to represent a yearly growth about 1.5 times higher than projected growth in national income.[4]

Budget and contract funding

There is some evidence that these increases in funding are going to be used to reverse what has been seen as a worrying trend in recent years, a falling proportion of science expenditure devoted to fundamental research. Fundamental research has been hit both in the Academy sector, which concentrates in that area, and in the branch sector, to which the vast bulk of scientific funding, about 80 per cent, is allocated (with the Academy and higher educational establishments (*vuzy*) taking about 10 per cent each).[5] Because of a squeeze in budgetary funding the Academy, centrally but even more noticeably in the republics, has been forced to derive an increasing proportion of its funds from contract research, i.e. research projects based on agreements made directly with users. More details, including details on the effects on fundamental research, are provided in Chapter 3.

The same trend of an increasing share of funding coming from contract sources that has been seen in the Academy in recent decades has also been evident in the branch and *vuz* sectors. Nedil'ko states that in a sample of 49 branch institutes in Minsk the percentage of funding derived from contracts increased between 1976 and 1981 from 40.7 to 53.5, with the percentages for the industrial institutes in the sample being 42.1 and 61.6.[6] Enterprise financing of branch research (which is essentially contract financing) rose for the country as a whole from 40.6 per cent in 1965 to 61.4 per cent in 1980, with the 1980 figure for industrial R&D being 93 per cent.[7] This has not left a lot of room for fundamental research in the branch sector. Efforts are being made to increase the share, with the Academy increasing its supervision of fundamental research in major branch institutes.[8]

The best example of the increase in contract work in *vuzy* comes from the Belorussian republic. In 1960 contract work worth 557,800 rubles had been done. By 1979 the figure had reached nearly 26 million rubles. This represented over 85 per cent of total allocations to *vuzy* for R&D. They made 1.8 million rubles clear profit from contract research between 1971 and 1975, and then 4.7 million in 1978 alone.[9]

Table 1.1 Funding of science in the USSR

Year	*Science budget (billion rubles)*	*Science budget as % of total budget*	*Yearly growth*	*Total expenditure (including capital investment, billion rubles)*	*Yearly growth*	*Total expenditure as % of national income*[a]
1940	0.105	0.6		0.3		
1950	0.539	1.3		1.0		
1960	2.172	3.0		3.9		
1970	6.425	4.2	11.1	11.7		4.04
1971	6.916	4.2	5.7	13.0	11.1	4.26
1972	7.301	4.2	5.5	14.4	10.7	4.59
1973	7.502	4.0	2.7	15.7	9.0	4.65
1974	7.899	4.0	5.2	16.5	5.1	4.66
1975	7.893	3.6	–0.07	17.4	5.5	4.79
1976	7.897	3.4	0.05	17.7	1.7	4.59
1977	8.189	3.3	3.6	18.3	3.4	4.51
1978	8.777	3.3	7.1	19.3	5.5	4.53
1979	9.280	3.3	5.7	20.2	4.7	4.58
1980	9.946	3.3	7.1	22.3	10.3[b]	4.86
1981	10.710	3.5	7.7	23.4	4.9	4.81
1982	11.553	3.4	7.9	24.9	6.4	4.76
1983	12.541	3.5	8.6	26.4	6.0	4.82
1984	13.015	3.5	3.8	27.8	5.3	4.88
1985	13.446	3.5	3.3	28.6	2.9	4.95
1986	14.327	3.4	6.6	29.5	3.1	5.0
1987	[16.451][c]		[14.8]	[30.8]	[4.4]	
1988	[17.800]		[8.2]	[32.9]	[6.8]	

Sources: *Narodnoe khoziaistvo SSSR*, various years; *Pravda* 18 November 1986, p.3; *Izvestiia* 20 October 1987, p.3

Notes: a. I have used official Soviet figures for Net Material Product. The US National Science Foundation uses its own estimate of Soviet GNP for comparison with US science shares of GNP. These suggest a reduced Soviet share of roughly 0.8 per cent.

b. This unusually high increase seems to be an artifice deriving from an upward revision of data from 1980 on. The original figure for 1980, given in *Narodnoe khoziaistvo SSSR v 1980 godu*, was 21.3. Later year books give 22.3, but without providing any figures for immediately preceding years.

c. Figures in brackets are projections.

Although many recognise the benefits of contract research in tying research closer to the needs of users and in providing considerable funds for institutes, there is also concern that too much contract research is in fact made up of small 'firefighting' tasks that bring very little economic or scientific return. In all sectors of science there is concern over *melkotem'e* (the excessive concentration on minor tasks), and the growth in contract funding has been seen as one of the major culprits. (On average centrally financed projects are five times larger in cost terms than contract projects.[10]) In the last decade or so this produced a concern to strengthen the central control of scientific funding, through 'programme-goal planning' and the centralisation of branch R&D financing in central ministry funds. Under Gorbachev, however, with the 'commoditisation' of science, attention is turning again – albeit ambiguously – to contract research. The details and implications of these changes will be discussed further in later chapters.

The question of contract research is particularly sensitive as far as the distribution of scientific funding between fundamental and applied research is concerned. Many commentators also consider that development is underfunded compared with applied research. This is the statistic which varies most in Soviet sources, but it is most often said that the split between fundamental, applied and developmental work is 10:60:30, whereas a proportion 10:30:60 is considered more appropriate. The excessive expenditure on applied research is because economic managers find it easier to fulfil research plan indicators than development and implementation indicators (since the latter require all the incentive barriers to technological innovation that exist in the USSR to be overcome). Thus R&D expenditure plans are overfulfilled for their research component, but underfulfilled for their development and implementation component.[11]

Capital and current expenditure

While institutes' receipts are divided between budgetary allocations and returns from contract research, their outlays are in the first instance divided between capital and current expenditures. Direct figures for capital expenditure on science are not given in Soviet statistics. However, the figures provided in *Narodnoe khoziaistvo SSSR*, the statistical handbook, on total science expenditure (Table 1.1) apparently do include capital investment. The figure provided for 1985 is 28.6 billion rubles, while Ryzhkov, speaking at the presentation of the Twelfth Five-Year Plan, gave for the same year, with capital investment explicitly excluded, a figure of 24.8 billion. This suggests capital investment in science of 3.8 billion rubles in that year, or 13.3 per cent of total expenditure. It is said in one source that the per annum growth in capital investment between 1971 and 1980 was 6 per cent,[12] which is slightly below the 6.2 per cent per annum increase in total science expenditure over the same period. Ryzhkov is claiming a 70 per cent increase in capital investment by 1990, which is somewhat higher than the increase in science expenditure as a whole.[13] There is particular emphasis on the Academy of Sciences. In mid-1986 it was said that capital investment in the Academy was growing at 1.7 times the rate achieved in the Eleventh Five-Year Plan (1981–5), a significantly greater rate than for capital investment in the economy as a whole.[14] That higher rate of growth for the science sector is set to continue for the foreseeable future.[15]

The capital expenditure budget has traditionally been devoted to the construction of buildings and fixed capital, in the same inefficient way that has dogged all sectors of the Soviet economy. (Capital construction plans are usually seriously underfulfilled. For example, the Academy's plan for 1980 was only 89 per cent fulfilled.[16]) But in recent years an effort has been made to devote more capital expenditure to building up experimental-prototype facilities and the purchase of equipment. In the 1980s about 40 per cent of total capital expenditure in science went on developing such facilities.[17] In the current Twelfth Five-Year Plan that percentage is set at no less than 50 per cent.[18] The emphasis on experimental-prototype and developmental facilities has become increasingly evident in the case of the Academy. During the current Twelfth Five-Year Plan capital investments in this area are to increase two and a half times over the previous plan, taking 15 per cent of total capital investment, up from 10.4 per cent in the Eleventh and 4.1 per cent in the Tenth Five-Year Plans.[19] The Academy percentage is still considerably less than that for the science sector as a whole, a reflection of the predominantly applied R&D role of the branch sector.

As far as equipment is concerned, in 1959 24 per cent of capital expenditure in the science sector as a whole went on new equipment, up to 40 per cent in 1970. In 1980 about 65 per cent of the increase in capital investment was devoted to equipment.[20]

These diversions of capital funds into developmental facilities and equipment have meant a considerable squeeze on the construction of new buildings, both research premises and living accommodation for researchers (see Chapter 9). But despite this some commentators consider that institutes are still expected to fund too much of their equipment purchases from current costs.[21]

Current costs have traditionally gone predominantly on wages, although a determined effort has been made to reduce their share in recent decades. Between 1950 and 1975 the proportion of science spending devoted to wages declined from an extraordinary 80 per cent to 43 per cent,[22] with the decline continuing. It is a development that has not been without its problems. Wages in science have declined relative to wages in other sectors of the economy, producing some problems in attracting the best young personnel. Further, in an effort to maintain research personnel at a time of pressure on overall wage allocations, research managers have drastically reduced the numbers of ancillary personnel in the administrative, technical, and research assistance areas.

To some extent the savings on wages have been used to increase purchases of equipment out of current cost budgets, but the main increase seems to have been in the 'overheads' category. One source gives a breakdown of current costs in the 1970s which shows that 49–50 per cent went on wages, about 20–30 per cent, depending on the type of institute, on overheads, and about 20 per cent on supplies, fuel and power, and payments to outside organisations. This source then states that only 10–15 per cent of expenditure went on the 'development of the material base of science', i.e. the purchase of special equipment and instruments. This would seem to mean that 10–15 per cent of current costs went on non-capital equipment.[23] A recent breakdown of the budget of the Lithuanian Academy showed 50 per cent going on wages, 25 per cent on 'the roof over our head, light and heat', and the rest, about 1,500 rubles per scientist per year, on equipment, instruments, and materials.[24]

Equipment expenditures

The combination of the increases in expenditure on equipment from both capital and current budgets has produced significant increases in aggregate spending on equipment. It leapt from a 3.8 per cent per annum growth rate in 1963 to 16 per cent in 1965, while between 1965 and 1970 a per annum growth rate of 12.7 per cent was maintained. In 1966 7 per cent of total science expenditure went on equipment. Between 1971 and 1976 supply of materials moved from 15 to 17 per cent of total scientific expenditure, with about 60 per cent of this going on instruments and laboratory equipment.[25] These increases in expenditure on equipment led to substantial increases in the amount of fixed capital per worker (*fondooruzhennost'*), particularly in the 1960s. Between 1960 and 1968 there was a 291.4 per cent increase in this indicator for the science sector as a whole. However, a decline in growth set in in the late 1960s. The indicator grew at over 12 per cent per annum between 1961 and 1965, but declined to 11 per cent between 1966 and 1970 and about 8 per cent from 1971 to 1973.[26] In 1987 the late academician Anchishkin stated that in branch science *fondooruzhennost'* was less than half the figure for the average industrial worker, whereas in the USA the figure for scientific workers was at least one and a half times that of an industrial worker.[27]

There was a movement to expand the Academy's equipment-making capacity from the mid-1960s, along with the general campaign to expand that of the country as a whole. Thus between 1965 and 1971 the Academy's fixed capital doubled while the value of its equipment and apparatus stock increased by nearly 250 per cent.[28] During the Ninth

Five-Year Plan (1971–5) small-run production of instruments within the Academy increased by 430 per cent, while between 1965 and 1975 planned expenditure on equipment within the central Academy increased by 838 per cent (as against a 238 per cent increase in wages).[29] This was a time during which there was also a readiness to import equipment from abroad. Academy purchases of scientific equipment from capitalist countries quintupled between the Eighth and Ninth Five-Year Plans, while purchases from socialist countries trebled.[30]

But foreign currency problems, the lobbying of the new anti-imports president of the Academy, A.P. Aleksandrov, and then, finally, Western embargoes at the end of the 1970s removed this option. Consequently, with the decision being made that the Academy would have to set up its own dedicated instrument-making capacity, in 1977 the Scientific–Technical Association (NTO) of the Academy was established.[31] It is said that the replacement of the original general director following the April 1985 Central Committee plenum was instrumental in the enormous growth experienced by the NTO, particularly in terms of its production capacity. In mid-1987 its Special Design Bureau alone employed nearly 3,000 people. In the current Five-Year Plan plans have been approved for the construction of 11 new factories, of which on five work had already started in the first half of 1987. It is planned to double the NTO's output of scientific instruments in the current Five-Year Plan and increase the production of instruments throughout the Academy as a whole by a factor of two and a half. The latter increase entails producing instruments valued at 100 million rubles per annum by 1990, and devoting 15 per cent of total capital investment in Academy science to the expansion of experimental and instrument-making capacity in the Twelfth Five-Year Plan (against 10.4 per cent in the Eleventh and 4.1 per cent in the Tenth).[32]

Summary

To summarise the funding situation, despite a levelling off over the last 15 years, funding for science is still at relatively generous levels. While a large amount of money is wasted on low-priority and non-productive research, new fields which are given priority (usually a little late) can still attract lavish attention. The Academy receives a small proportion of total funds available, and considers that its research suffers as a result. There is a widespread feeling that there is considerable wastage in the branch sector, to which the vast bulk of total funding is allocated. Traditional deficiencies in funding for equipment are being redressed, to the cost of wages and capital construction. There is no sign that science will be a major victim of any financial squeeze or investment reallocations that might be applied in the future.

Personnel

Soviet data on numbers of research personnel are no more compatible with Western data, even in purely quantitative terms, than data on funding. The Soviets firstly provide statistics on the numbers of personnel employed in the 'science and scientific services' sector of the economy. Not only does this statistic cover a different range of institutions than would usually be seen as research organisations in the West, but it includes all personnel working in those institutions regardless of their function. The other statistic that the Soviets provide is the number of 'scientific workers' (*nauchnye sotrudniki*). This category includes all people engaged in research work within research institutions (broadly defined), plus all people with advanced degrees regardless of where they work. It also includes 'scientific–pedagogical personnel', i.e. teachers in higher educational institutions, whereas the 'science and scientific services' category does not. A number of Western analysts have wrestled with these statistics in an attempt to arrive at comparable data. The general conclusion is that the USSR has considerably more people working in research than in any Western country, both overall and as a percentage of the working population.[33]

The USSR maintains its advantage despite a marked slowdown in recruitment into research work in the Soviet Union in recent years. Table 1.2 shows the decline from 1975 to 1976, with a further drop in the early 1980s. The emphasis is still very much on cutting the deadwood out of the scientific sector, and we can therefore expect to see further decreases in growth rates of scientific personnel. The cuts seem to have been experienced by all sectors, although most heavily by the 'science and scientific services' category. This is because the reductions have been felt most strongly among ancillary and administrative staff, and less strongly among scientific–pedagogical personnel (who are not included in the 'science and scientific services' category).[34]

A large proportion of personnel employed in the science sector are not classified as scientific workers. Since the 'scientific worker' and 'science and scientific services' categories are not comparable, we cannot arrive simply at a proportion. However, virtually all sources suggest that only about 25 per cent of those working in Soviet science are 'scientific workers'. This was the global figure given by Duzhenkov in 1978, and the one arrived at in a 1980 survey of 26 research organisations in 20 branches of the economy.[35] But obviously there is going to be variation from sector to sector and from institute to institute. Thus for the republican Academies of Science the percentage ranges from 50 per cent for the Georgian Academy to 18 per cent for the Ukrainian Academy. The All-Union Academy scored 25.4 per cent in 1975, while the Siberian Academy scored 18.2 per cent.[36] The differences between the various Academies are explained by the varying amount of developmental and even production work that each does. Both the Ukrainian and Siberian Academies have major developmental and production facilities, and therefore have large numbers of staff who are classified as 'production workers'. The importance of 'production workers' in the Siberian Academy can be seen in the breakdown of staff for 1976 given in Table 1.3. The category 'scientific–technical workers' seems to cover laboratory technicians, assistants, etc., a category which is generally considered to be underrepresented in Soviet science.

Table 1.2 Scientific personnel in the USSR

Year	*Science and science service personnel (000s)*	*Yearly increase (%)*	*Scientific workers (000s)*	*Yearly increase (%)*
1940	362			
1950	714		162.5	
1960	1,592		354.2	
1965	2,403		664.6	
1970	2,999		927.7	
1971	3,374	12.5	1,002.9	8.1
1972	3,544	5.0	1,056.0	5.3
1973	3,735	5.4	1,108.5	5.0
1974	3,864	3.5	1,169.7	5.5
1975	3,790	−1.9	1,223.4	4.6
1976	3,860	1.8	1,253.4	2.5
1977	3,969	2.8	1,279.6	2.1
1978	4,069	2.5	1,314.0	2.6
1979	4,264	4.8	1,340.6	2.0
1980	4,379	2.7	1,373.3	2.4
1981	4,477	2.2	1,411.2	2.8
1982	4,475	−0.04	1,431.7	1.5
1983	4,471	−0.1	1,440.0	0.6
1984	4,508	0.8	1,463.8	1.7
1985	4,554	1.0	1,491.3	1.9
1986	4,546	−0.2	1,500.5	0.6
1987			1,517.9	1.2

Source: *Narodnoe khoziaistvo SSSR*, various years.

Strangely the percentages of research workers in the branch sector seem to be higher than in the Academy. For example, the institutes in the All-Union Scientific–Production Association for the Paper Industry (VNPObumprom) had 2,571 staff in 1983, of whom 1,537 (59.8 per cent) were research workers. In 1984 in the Ministry of Mineral Fertilisers 5,500 of about 13,000 people involved in R&D were research workers.[37] These higher percentages are presumably an indication of just how little development and prototype production work is done in branch institutes, as well as of the serious shortage of ancillary staff in those institutes.

Table 1.3 The distribution of staff of the Siberian Academy of Sciences, 1976

	Numbers	*%*
Scientific workers	6,282	18.0
Scientific–technical workers	9,884	28.3
Production	13,167	37.6
Administration	2,607	7.5
Service	1,983	5.7
Total	33,923	

Source: A.K. Romanov, L.A. Androsova, and A.F. Felinger *Nauchnye kadry Sibirskogo otdeleniia AN SSSR. Metody: resul'taty statisticheskogo issledovaniia*, Nauka, Novosibirsk, 1979, p. 9

Table 1.3 shows that 7.5 per cent of the staff of the Siberian Academy of Sciences in 1976 were administrative workers. This figure has also been given for the institutes of the Academy of Science as a whole, while that for branch institutes is said to be a little higher.[38] Some commentators have expressed concern at the tight restrictions on the employment of administrative personnel, and the need to conceal them by falsely describing them as research personnel in statistical reports.[39]

As can be seen from Table 1.4 a considerable proportion of Soviet scientists are concentrated in the branch network, i.e. the research organ- isations subordinate to the numerous ministries supervising industrial and agricultural production. Figures vary considerably from slightly under 50 per cent to as high as 70 per cent of scientists working in branch institutions.[40] One presumably needs to be more precise in the treatment of scientific–pedagogical personnel in order to arrive at a universally accepted figure. Whatever the figure, very few scientific workers, less than 10 per cent, work in enterprises, although in some industries, particularly agricultural and heavy machine construction, the automobile industry, and the machine tool industry, the percentages can be far higher, from 25 to 50 per cent.[41] The great majority of these branch R&D personnel are engineers, and concern is often expressed that their qualifications are inadequate.

The figures for the Academy of Science are quite straightforward. The relatively low percentage of Soviet scientists working in the Academy networks is noteworthy. Table 1.5 shows the breakdown of staff for the central Academy, the union republic Academies and the branch Academies.

Table 1.4 The distribution of scientific workers by sector and discipline

Year	*Total scientific workers (000s)*	*In Academy institutions*	*In branch institutions*	*In vuzy*	*In physical, life and earth sciences*	*In technical sciences*
1960	354.2				85.3	129.8
1970	927.7	85.1	420.0	422.6	205.9	409.5
1980	1,373.3	125.6	608.3	639.4	289.8	635.2
1985	1,491.3	137.3	662.2	691.8	305.1	708.2
1986	1,500.5	141.9	661.1	697.5	304.5	715.7
1987	1,517.9				310.4	719.4

Source: *Narodnoe khoziaistvo za SSSR 70 let*, pp. 62–3; *Vestnik statistiki* no. 7 (1988), pp. 65–6

Table 1.5 Scientific workers in the Academies

Year	*Soviet Academy of Sciences*	*Union-republic Academies*	*Branch Academies*[a]	*Total*
1970	35,115	36,175	14,621	85,911
1975	41,836	44,361	18,343	104,540
1976	42,951	45,458	19,267	107,676
1977	44,168	45,899	21,663	111,730
1978	45,721	46,824	21,683	114,228
1979	46,115	47,344	27,843	121,302
1980	47,825	49,079	28,968	125,872
1981	49,019	50,473	29,145	128,637
1982	49,523	50,443	28,859	128,825
1983	50,711	50,749	28,243	129,703
1984	52,100	51,740	28,778	132,618
1985	55,583	52,791	29,158	137,532
1986	58,492	53,988	29,710	142,190
1987	60,518	55,231	30,613	146,362

Source: *Narodnoe khoziaistvo SSSR*, various years; *Vestnik statistiki* no. 7 (1988), p. 67
Note: a. The branch Academies are the Lenin Academy of Agricultural Sciences (VASKhNIL), and the Academies of Arts, Medical Sciences, Pedagogical Sciences, and Communal Economics. VASKhNIL is by far the largest, representing roughly 75 per cent of the branch Academy category.

The final major category is higher educational institutions (*vuzy*). The situation here is complicated by the fact that, as in the West, most *vuz* staff have teaching commitments. Analysts have used various approaches to reaching full-time research equivalents for these staff.[42] Some percentage of *vuz* staff are in fact full-time researchers, since they work in special research institutes and laboratories attached to universities. As Table 1.6 shows, this percentage increased rapidly throughout the 1960s and 1970s but was still less than 20 per cent of total scientific–pedagogical personnel.

Table 1.6 Numbers of scientific and scientific–pedagogical workers in *vuzy*

Year	*Total*	*Scientific–pedagogical workers*	*Scientific workers*	*Scientific workers as % of total*
1950	86,542	84,472	2,070	2.4
1955	119,059	116,381	2,678	2.2
1960	146,915	141,138	5,777	3.9
1965	221,800	204,800	17,000	7.7
1970	348,872	303,255	45,617	13.1
1974	410,818	340,322	70,496	17.2
1976	441,500	362,300	79,200	17.9
1980	493,900	334,700	83,400	16.9
1985	516,500	345,600	93,400	18.1
1986	517,400	338,900	97,000	18.7
1987	528,700	345,200	99,400	18.8

Sources: The 1950–76 figures are taken from L.E. Nolting and M. Feshbach, 'R&D employment in the USSR – definitions, statistics and comparisons', in *Soviet Economy at a Time of Change*, vol. 1, Joint Economic Committee, US Congress, Washington, DC, October 1979, p. 733. The 1980–7 figures are from *Narodnoe khoziaistvo SSSR za 70 let*, p. 63, and *Vestnik statistiki* no. 7 (1988), p.66

There is a small 'others' category made up of research workers in libraries, museums, and administrative organisations. The fact that all people with advanced degrees are counted as scientific workers means that administrative organisations employ people classified as scientific workers although they do no research work. In Georgia in 1982 such 'scientific workers' made up 5 per cent of the total number.[43] A.I. Shcherbakov's figures show 2 per cent for the whole of the Soviet Union and 0.6 per cent for Siberia.[44] One Soviet analyst attempts to show how the number of non-scientific 'scientific workers' is increasing (and uses a critical tone to do so) by looking at the difference between the number of people with degrees and the number with academic titles (such as professor, senior research worker, etc.). In 1965 the gap was 1.6 per cent. By 1980 it had climbed to 13.7 per cent. According to this analyst's figures, of the approximately 100,000 advanced degrees awarded between 1975 and 1978, about 5,000 were given to people who did not work in academic institutions.[45]

Summary

Arguments are evident in the Soviet and Western sources about the precision of Soviet statistics on research personnel. Despite this, and cuts in recruitment in recent decades, the size of the research community is impressive, and when the funding devoted to science is also considered we are left in no doubt as to the commitment of the Soviet leadership to an R&D sector of international significance. What now needs to be considered is whether the leadership is getting the return on its investment that it expects.

The return on investment

Over the years a considerable amount of work has been done by Western analysts on the quality and effectiveness of Soviet research, in both the fundamental and applied spheres. It is recognised, however, that the work is of a scattered and fragmentary nature, and that there are serious methodological problems involved in all the various approaches adopted. However, the high degree of agreement between the findings of all studies is striking, and it is an agreement which is not favourable to the Soviet Union. There is some debate over some areas of work, and even general unanimity on Soviet excellence in particular fields. But the general impression is of a level of performance lagging considerably behind that in the West.

Fundamental science

In the fundamental sciences the main methods used have been publication and citation analyses and records of the impressions of *émigrés* from the Soviet Union and of Western scientists familiar with the work of their Soviet counterparts, usually as a result of academic exchanges or collaborations.

Publication and citation analysis has been used perhaps least of all. Presumably this is because it is an excessively time-consuming job. But there are also considerable methodological problems involved. There are technical problems of the coverage of the citation indexes, but perhaps more importantly publication and citation practices are so 'culture specific' that the results of this type of analysis require very considerable interpretation. Furthermore, of course, classified research, which many consider to be among the Soviet Union's best, will not feature at all. Irvine and Martin recognise this in their study of high-energy physics, and so combine their citation and publication analysis with the more subjective 'peer evaluation' approach. Their findings overall are of poor perfomance at Soviet accelerator facilities, including in terms of publication and citation results.[46] Narin, Frame and Carpenter, in a number of studies covering all the major natural science disciplines, found consistently low levels of citation of Soviet articles, even in areas in which the Soviets are generally acknowledged to do good work. The latter fact is enough to make the authors wary of drawing any conclusions from their data on the quality of Soviet research. They are even reluctant to declare categorically, although they are tempted to do so, that Soviet science is highly isolated from the international mainstream.[47]

Enormous programmes for interviewing Soviet *émigrés* have been undertaken in recent years, following the large-scale primarily Jewish emigration of the 1970s and 1980s. However, relatively few results in the field of science have appeared, particularly with regard to evaluation of Soviet research. The Harvard-based project 'Soviet R&D: information and insights from the third emigration' concentrated on *émigrés* with R&D experience, and some of its results have now appeared.[48] The survey investigators determined that much of the evidence presented on the quality of Soviet science 'is in the realm of opinion rather than data based on direct experience'. Nevertheless the respondents 'strongly reinforced [the] judgement of relative Soviet backwardness', although the characterisation is applied particularly to applied technology.[49] Beyond responses to structured surveys *émigrés* have provided a considerable input into our understanding of Soviet science through their own writings. These accounts range from the damning to the respectful. The best example of the former is Mark Popovsky's book.[50] Although he makes no claims to an analysis of the quality of Soviet science, his picture of corruption and demoralisation hardly allows for either high quality or high productivity. Most of his compatriots publishing in the West, while not always as uncompromising and indeed at times concerned to point out to the West the underlying potential the Soviets have, nevertheless confirm his general picture.[51]

The final source of information we have are the impressions of knowledgeable Western scientists. Many of these impressions are little more than 'throw-away' lines in the Western popular science press. An example is Peter Smith's characterisation of Soviet earth sciences as 'mediocrity that looks suspiciously like the result of isolation and inbreeding', with papers that are 'all too frequently vague, uninformative, impoverished in data and almost completely lacking in international outlook'.[52] When a greater number of Western scientists are brought together, as was done by the Scientific Affairs Division of NATO in September 1986, the conclusions are not surprisingly less black-and-white. Nevertheless the general picture of backwardness is confirmed, even in fields in which the Westerners felt it worthwhile and even essential to keep up with Soviet work or to work with Soviet scientists.[53]

By far the broadest surveys of Western scientists have been carried out under the auspices of the US National Academy of Sciences and based on questionnaires sent to American scientists going to the USSR on inter-Academy and science and technology exchanges, and to the hosts of Soviet exchange researchers. The so-called Kaysen Report is the most concerned with the fundamental sciences.[54] While the respondents were generally enthusiastic about the exchanges and about the scientific value of the cooperation in which they were involved, they still considered the USA to be ahead of the USSR in most scientific disciplines. Further, they tended to support the exchanges more for their political and cultural benefits than for purely scientific reasons.

All Western sources agree that in general Soviet fundamental research lags behind that of the West. There is also a considerable degree of agreement as to the reasons. *Emigré* writers tend to concentrate on the negative moral atmosphere in Soviet research produced by at best authoritarian and at worst corrupt managers, but there is also a surprising degree of consensus among even relatively well-disposed Western scientists that Soviet science suffers from old conservative managers wielding dictatorial powers. Some refer to funding and supply problems arising from the lack of priority given to fundamental research. All refer to equipment and ancillary service problems, usually the result of a technological lag in such fields as materials science and computing. All also refer to the negative effects on Soviet science of self-imposed barriers to international cooperation.

All this having been said, it should be pointed out that most Western sources are aware that the picture is far from uniform. It is recognised that the Soviets do far better in some areas than others, and that in some they are catching up faster than in others. Thane Gustafson points out that Soviet pure science is strongest in fields that depend least on material support, including mathematics as the most outstanding example.[55] It is considered that the Soviet educational system, while perhaps having other faults, does provide an excellent grounding in mathematics and the theoretical aspects of various disciplines.[56] It has usually been assumed that defence-related work is of higher quality, because of its higher priority, tighter quality control, and better management, although a number of analysts are now questioning that assumption (see Chapter 5).

It is noteworthy that in recent times Soviet commentators, including Soviet scientists, have confirmed the Western impressions of backwardness in the fundamental sciences.[57]

Applied science

It is generally accepted by Western analysts that Soviet applied research performs even worse than fundamental research. We have less direct access to Soviet applied institutes, but both *émigré* reports and those of foreigners with direct experience tend to support the finding.[58]

However, because applied research is seen as more susceptible to more-or-less precise evaluation by examination of its output – in terms of numbers of inventions and innovations, financial returns, and the quality of industrial products, as well as labour productivity – Western analysts have devoted considerable effort to detailed analyses of applied science and technological performance. These studies have been at both national and industry-by-industry levels.

Considerable attention has been devoted to sophisticated econometric techniques for estimating productivity levels and change in the Soviet industrial sector. Hanson and Pavitt have recently summarised this work, and came to the careful conclusion:

> The most that can be said in the Soviet case is that the relatively strong and prolonged deterioration of measured productivity growth makes it particularly likely that a slowdown in the rate of technological progress proper is involved.[59]

The difficulties involved in production function work have led to the adoption of other approaches. One has been to use Soviet data on technological indicators. The problems here are the vagueness of the data and their lack of comparability with Western statistics. However, Ronald Amann, for example, makes good use of such Soviet data. Particularly when admitting to a shrewd suspicion that the figures could be artificially inflated, he finds striking the continuous absolute decline since the early 1960s in the creation of prototypes of new machines and equipment, in *vnedrenie*, and in economic return indicators, with the declines especially evident in high-technology industries.[60] The consequences of these declines are seen in sluggish economic performance, poor productivity figures, and the virtual exclusion of Soviet exports of industrial and consumer goods from Western markets.[61]

Amann recognises the difficulties of all aggregated approaches to measurement, and consequently the 'Birmingham school', of which he is a key member, has adopted as its primary strategy a disaggregated approach, i.e. an industry-by-industry analysis. Their work, now in three volumes, covers a good range of Soviet industries,[62] and can be combined with the work of other scholars in the same and other areas.[63]

The original work of the Birmingham group, which was pretty damning of Soviet technology in a whole range of sectors, produced a 'revisionist' response from some Western analysts, including some from within the group, which suggested that some acknowledgement should be made of Soviet achievements, outstripping the West or narrowing the gap in some areas.[64] However, even with that taken into account sector-by-sector analyses tend to support the aggregate findings – that Soviet technology lags seriously behind the West.

As far as the level of technological development is concerned Soviet writers and political leaders are quite willing to support the Western findings. The Soviet press is full of complaints about poor technological performance by particular enterprises and industries. Soviet writers are also quite willing to cite the negative aggregate indicators that Western writers also use. There was a strong and publicised awareness as Gorbachev came to power that the indicators had worsened in previous years, particularly through the Eleventh Five-Year Plan. For example, in the Volgograd region yearly output of research in local institutes fell almost 20 per cent and the economic return on the work 35 per cent during that plan. It is claimed that, in the Ukraine for the Tenth and Eleventh Five-Year Plans, of 12 major science and technology indicators recorded seven showed severe declines in growth rates, four showed negative growth and one a small increase. (We are not told what the indicators are.[65]) Sectoral data presented in Soviet sources are not much better.[66] It is clear that more than anyone else the Soviet authorities themselves are disgusted by the return they get on their investment in applied research.[67]

The problem with all such information for our purposes is to what extent it says anything about the state of Soviet applied research. Many other factors could explain low labour and factor productivity, low quality and technological levels of products, and poor performance in world markets. Indeed most Western studies have concentrated on problems within industry and its managing organs, or in the transfer of work from institute to enterprise. It would be unfair to say that Western analysts assume that there are no problems in the research sector – it is rather that that sector receives considerably less attention. In this study I wish to avoid the usual concern with the three Vs – *vnedrenie* (introduction into production), *vedomstvennost'* (departmentalism), and *val* (gross output) – as explanations for poor technological performance, without in any way wanting to deny the importance, perhaps primary importance, of those issues. Here I am concerned essentially with research management, rather than industrial management.

On *a priori* grounds it might be possible to hypothesise that applied science would not suffer to the same degree from the problems of fundamental research – poor management and morale, low priority, and equipment problems. The applied sector gets the lion's share of funding and personnel resources, while the Soviet Union has an ideological preference for 'practical' over 'impractical' research. The Soviet Union also has a commitment to large-scale industry that should ensure priority for applied industrial research. In fact anyone with the slightest knowledge of Soviet science knows that such a hypothesis is entirely wrong. While the applied sector might do well in aggregate funding and personnel terms, the resources are spread thinly; the fundamental sector enjoys infinitely higher prestige, and management and morale problems in the applied sector are therefore probably worse; while the commitment to industry means a commitment above all else to gross output and consequently a very low priority for new technology. There is thus every reason to believe that applied research problems are even greater than in the fundamental sector.

The serious problems of the applied sector have received particular attention since Gorbachev came to power. An interesting feature of the traditional Soviet attitude to science–production relations was a tendency to assume that the scientist always knows best. This arose partly from the traditional prestige of science in the USSR and the impressive successes of some of their scientist-led programmes, and partly from a realistic recognition that because of a poor incentive structure the production sector could not be trusted with responsibility for innovation. It could be seen in the assumption that usually ruled in the Soviet press – a powerful body in such matters – that any ministry or enterprise avoiding proffered new technology must have been in the wrong.

Under Gorbachev that assumption has been questioned quite sharply. One of the features of the 27th Party Congress was the fierce criticism of the scientific community, particularly the applied sector (see the speeches of Gorbachev, Solomentsev, Pletneva, and Mysnichenko). The lead has been followed in the press, and we now find that enterprises are *not* automatically criticised for failing to welcome new technology, research institutes *are* criticised for sinking large amounts of money into projects despite the justified lack of interest on the part of users, and the problems of enterprises which have found themselves saddled with inadequate new technology are reported with sympathy.[68] At the same time the misbehaviour of research managers and the indolence of rank-and-file researchers have been reported with ever greater bluntness, while purges and closure of institutes have become common.

None of this allows us to allocate a precise degree of blame to the applied research sector for poor Soviet technological and economic performance. It does provide considerable data, however, for an analysis of the problems of Soviet applied research, particularly with regard to management and morale within the institutes. This will be described later.

Conclusion

We have seen in this chapter that the Soviet Union devotes very considerable resources to its fundamental and applied science. The investment has brought an impressive return. Soviet fundamental research is clearly an important and in some areas a leading part of the international scientific community. Its applied research has industrialised the country twice (in the 1930s and after World War Two, with it should be said some help from imported technology) and turned the USSR into a major economic power and military superpower. Having said that, the Soviets clearly are, and have reason to be, dissatisfied with the return on their investment. Their scientific institutes are experiencing serious problems, which means that they can barely maintain their position behind, much less catch up, the West in either fundamental or applied research.

Chapter two

Central party and governmental agencies

This chapter examines the top-level general policy-making bodies of the USSR, the Politburo and the Council of Ministers. The middle-level management bodies, the State Committee for Science and Technology (GKNT), and Gosplan are also included.

The Politburo

I have presented what little we know of the involvement in science policy making of the Politburo, the Soviet Communist Party's supreme body, in another publication.[1] We have no essentially new information since the appearance of that treatment in 1986. It is still evident from the brief summaries of the weekly Politburo meetings published in the press that it discusses a wide range of science issues. These discussions usually lead to published decrees of the party's Central Committee, often jointly with the Council of Ministers. The issues considered at this level include organisational structures, funding, and priorities. The Politburo also has, like all party bodies, its *nomenklatura*, i.e. a list of posts to which it makes appointments. The Politburo's *nomenklatura* is, hardly surprisingly, not published. We know that the president of the Academy of Sciences is on it, since Ligachev recommended Marchuk as the Politburo's candidate for president before his 1987 election.[2] We can be very confident that the Politburo also appoints the head of the other major science management body, GKNT. Below that we can only guess. It is highly probable that the Politburo at least vets the appointments of the heads of the major research programmes, such as the biotechnology programme. It seems less likely that the appointments of institute directors go that high.

The Politburo does not contain a direct representative of the science sector, and indeed responsibility for science is most probably split in the way it is for the entire party apparatus between control of the Academy and *vuzy* on the one hand and control of the branch R&D sector on the other. The former are considered to be part of the ideological sector, and as such are supervised by the Politburo member responsible for ideology, at present Vadim Medvedev. The chairman of the Council of Ministers, Nikolai Ryzhkov, a member of the Politburo, has overall responsibility for the branch R&D sector, along with the Central Committee secretaries in charge of industry.

The Politburo has a number of commissions. One Soviet *émigré* source claims that included among these commissions is one concerned with technology policy. However, the commissions appear not to be, as we might at first glance expect, subcommittees of the Politburo manned by members of the Politburo itself. Thus the commission for technology policy is apparently headed by the chairman of GKNT, with the head of the Central Committee's Science Department and the president of the Academy of Sciences as his deputies.[3] A 1979 *Kommunist* article referred in more general terms to special Politburo commissions, with scientists included in the membership, which work on the draft decisions of Central Committee plenums.[4] As far as we can tell the membership of these commissions does not differ greatly from the membership of other policy-advisory bodies, and so they can hardly be seen as an alternative source of information for the Politburo.

It appears that there is no formal post of scientific advisor to the General Secretary of the Party. Khrushchev writes in his memoirs that Kurchatov once suggested that it would be useful if Khrushchev were to appoint him as scientific advisor in his capacity as chairman of the Council of Ministers (a post Khrushchev held simultaneously with the General Secretaryship). The proposal came so soon before Kurchatov's death in early 1960 that it was never implemented.[5] There have been no other references to any other scientists occupying the post, although the physicist Evgenii Velikhov appears to have close links with Gorbachev. He has been described in the West as Gorbachev's science advisor,[6] but there is no evidence that he is one of Gorbachev's inner circle, either formally or informally.

We do not have much evidence that top leaders have very close personal links with scientists or adopt them as protégés. Direct personal contacts between scientists and top political leaders are not uncommon, however. It appears that in these contacts formal status or title is less important than personal qualities, and that two qualities are required of a scientist wanting to maintain and exploit such a relationship. The first is a willingness to provide scientific evidence to support the politician in policy disputes at the top level. Brezhnev describes in his memoirs *Tselina* (Virgin Land) how on the recommendation of D.M. Kunaev he appointed the director of the Kazakh Institute of Soil Science as his agricultural consultant. In his struggle with the Ministry of Agriculture over agricultural policy in the Virgin Lands he used the probably not altogether objective findings of 'his' scientist to support his case.[7] The second required quality is that the scientist possess undoubted patriotic loyalty, in Khrushchev's case best demonstrated by a willingness to engage in defence research. Khrushchev quite clearly did not trust Petr Kapitsa because of his unorthodox attitude to defence work, a lack of trust which led to, at the very least, Kapitsa's being refused permission to travel abroad and being refused funds for a particular project. (This refers only to Kapitsa's difficulties under Khrushchev, not his even greater problems under Stalin.[8]) Those on whom Khrushchev relied the most had strong military backgrounds, for example, the two Keldyshs, Kurchatov and Lavrent'ev. Given the increasing presence of defence scientists in the Academy of Sciences (see Chapter 3) and the increasingly public role of scientists of defence background, such as Raushenbakh, there seems little reason to believe that this qualification has been removed.

What are the benefits of such a relationship for the scientist? They are not hard to find. In any political system access is a source of enormous power, and in none more so than a closed bureaucratic system such as the Soviet Union. Certainly scientists were able to take their ideas, and the ideas of their colleagues, to Khrushchev, with Lavrent'ev's conversion of Khrushchev to the idea of opening a branch of the Academy in Siberia a good example.[9] Also they were given great power within the scientific community – to determine which issues would be taken to the top authorities, to determine whether approaches by other scientists to the authorities would meet with success or not (according to Khrushchev it was on Kurchatov's advice that he decided not to fund one of Kapitsa's projects),[10] and the allocation of resources even when the issue does not go to the top authorities. Thus Kurchatov had enormous power to determine the distribution of resources within the physics community. For example, it seems to have been his decision that some resources be moved from the accelerator programme to cosmic ray research in the mid-1950s, following an earlier rejection of such a proposal.[11] Despite the undoubted power of these people, it should not be thought that they have any guarantee of success in their approaches to the political authorities. For example, while Lavrent'ev was successful in getting the Siberian division of the Academy, he was not able to persuade Khrushchev to fund a Far Eastern division.[12] Certainly one should not exaggerate the political status or influence of these high-level advisors. They are essentially advisors, not colleagues, and they are not in a position to dictate to the political authorities. However, they are highly important people within and for the scientific community. They tend to operate outside the formal institutional scientific hierarchy, and indeed could well be more important people than the heads of those formal hierarchies.

There are other scientists who, although they do not have regular access to political leaders, in the sense that they can readily see such leaders on their own initiative, are used by top-level political bodies in the policy-making process. Thus Iurii Ovchinnikov, the late head of the Soviet microbiology programme, was called to present a report to the Politburo on biotechnology. In his words he 'participated in the discussion of the long-term outlook of the development of biotechnology'.[13] Clearly these scientists have the opportunity to influence policy-making. However, their activities would be much more constrained by the framework established by the political authorities than those who have the right of personal access.

When the Politburo does interest itself in an issue, one can expect a strong and immediate reaction. A.A. Baev, the academician–secretary of the Academy of Science's Biochemistry Department, described the 1974 Central Committee and Council of Ministers (CC-CofM) decree on molecular biology as one of the most important organisational measures of recent years, which 'as it were removed the brakes from the development of physico-chemical biology and gave it enormous moral force'.[14]

However, such a positive reaction cannot be guaranteed. Baev gave no details on the precise consequences of the decree, but he most probably had in mind increased funding allocations and priority in construction and supply schedules. However, since 1974 a string of decrees on the microbiology industry have been issued and the topic has been discussed in the Politburo on a number of occasions, in the context of general dissatisfaction with Soviet performance in the field.[15] As another example, in 1985 in an interview in *Sovetskaia Rossiia* with a leading proponent of rotor technology, details were given of continuing inertia in the implementation of the new technology despite, as was perhaps pointedly remarked, the September 1984 Politburo discussion of the issue (with an unpublished decree resulting). In May 1986, following an article on the lack of diffusion of the new technology which appeared in the press in April 1986, the Politburo had again to discuss the matter and criticise some ministries for their continuing lack of enthusiasm.[16]

Central Committee apparatus

The Politburo is assisted in its involvement in science policy-making by the Central Committee apparatus. In my 1986 publication I described in considerable detail the work and functions of the Central Committee apparatus, and I do not intend to repeat the treatment here. But some of the more important features of the apparatus deserve repeating. In 1986 I summarised its basic functions as follows:

> There are two sides to the Central Committee apparatus. It works, firstly, as a giant staff office serving the Politburo and Secretariat. As such it has two essential functions. It controls access to the executive organs it is serving, but at the same time ensuring that those with genuine business get the access required. It also prepares documents for the Politburo and Secretariat – summaries of submissions and memoranda, position papers, and draft decisions and 'directive documents'.
>
> But it has, in addition, an important control function. Like all party organs it has major personnel powers, summed up in its *nomenklatura*, and is also charged with ensuring the fulfilment of policies set by the Politburo. It is expected to exercise these functions from both a political and an economic or technical standpoint.[17]

To fulfil these functions the apparatus, at least until recently, had a Department of Science and Educational Institutions. All the evidence suggests that this Department is on the ideological side of the apparatus, and is responsible only for the Academy of Sciences and the social sciences in general. Those staff we can identify have almost exclusively ideological backgrounds, although its most recent head, V. A. Grigor'ev, after working in the Central Committee apparatus between 1966 and 1972, from 1976 to 1985 was rector of the Moscow Energy Institute, a major teaching institute. He was elected a corresponding member of the Academy of Sciences, with a specialisation in low-temperature heat exchange, in 1981. It appears that under the latest 1988 reorganisation of the Central Committee apparatus the Department of Science has been merged with the Propaganda and Culture Departments into an Ideology Department headed by A.S. Kapto, a professional party apparatchik.

It appears that branch R&D is not supervised by the Science Department but rather by the relevant branch department of the apparatus. These branch departments have included the Departments for Agriculture and Food Industry, Chemical Industry, Defence Industry, Heavy Industry and Power Engineering, Light Industry and Consumer Goods, and Machine Building. It appears that under the new arrangements the branch departments have been abolished or, more probably, merged into a large Socio-Economic Commission or Department. One strongly suspects that the new super-department will have a branch structure not dissimilar to the old departments.

Scientists are quite well represented on the new Central Committee commissions. These are new bodies made up of senior executives from all sectors of society which are supposed to exercise policy-making guidance over the full-time apparatchiki of the Central Committee departments. Marchuk, the president of the Academy of Sciences, is a member of the Ideological Commission, which presumably has responsibility for the Academy. K.V. Frolov, the Academy's engineering vice-president, is a member of the Socio-Economic Policy Commission, vice-president E.P. Velikhov is a member of the International Policy Commission, and vice-president A.A. Logunov is a member of the Legal Policy Commission.[18]

It appears that the Central Committee apparatus is an important channel for scientists who want to plug into the policy-making process, with documents, memoranda, position papers, proposals, and simple letters for senior party leaders, at Politburo level in particular, having to go through the apparatus. What is not clear is to what extent the apparatus is simply a post-box, playing essentially a secretarial role, or whether it exercises any influence over the policy-making process through its treatment of the documents it receives. It does employ outside consultants, and some of these are highly controversial scientists, for example the geneticist Nikolai Dubinin, who would almost certainly try to push a particular line.[19] But in a way this fact represents an argument for the view that the apparatus is not seriously involved in policy-making. Its own staff would not appear to have the scientific knowledge, much less the standing, to 'take on' the scientific élite or even to take sides. Further, the scientific community apparently has other routes to the top, both formal and in particular informal. The indications are that serious consideration of policy options takes place either in direct contacts between senior scientists and top party leaders or within the state apparatus.

The role of the Central Committee apparatus is much more evident in the area of what might be called 'ideo-political control'. Firstly there is control of personnel appointments. The evidence suggests that the apparatus includes on its *nomenklatura* the directors and deputies, as well as middle-management, of major institutes. We know too little about how the system works to come to any firm conclusions, but it appears probable that, while the apparatus does not have the capacity and possibly the desire to make technical judgements on the merits of various candidates, it will certainly make political and bureaucratic judgements.

Perhaps even more sinisterly the Central Committee apparatus is heavily involved in the control of conferences, travel, and publishing, i.e. who should get exit visas and who should be allowed to publish. In his *émigré* book *Nomenklatura* Mikhail Voslensky has included a chapter 'One day in the life of Denis Ivanovich', which is a fictional although well-informed account of the working day of an official in the Central Committee's Science Department. Typical of the issues with which he is concerned are who should be allowed to go on an overseas delegation and which authors should be struck off the list of next year's publications. The victims of his labours tend to be Jews and those with 'doubtful' backgrounds.[20] It could be said that these political controls, while unpleasant for the victims and undoubtedly having some unmeasurable negative effect on Soviet scientific performance, are not a vital part of the Soviet science policy-making process. They cannot and must not be ignored or discounted, but they will not be a major feature of this study.[21]

There is no reason to believe that science policy-making is any different from any other aspect of policy-making in the Soviet Union, in that the top party bodies, in particular the Politburo, are ultimately responsible for and closely involved in the taking of all major decisions. However, not being an essentially political or ideological activity, science is formally and to a considerable extent practically managed in the state sector, and we must now turn our attention there.

Council of Ministers

The Council of Ministers is formally the supreme executive body of the USSR, roughly equivalent to a Western Cabinet. Its membership includes, as of November 1988,[22] a chairman, at present Nikolai Ryzhkov, often described in the West as the Soviet Prime Minister; two first deputy chairmen (both of whom occupy other posts as chairmen of the Agroindustrial Committee and of Gosplan); 11 deputy chairman, also occupying other positions, including the chair of GKNT; 87 ministers and chairmen of State Committees, plus some deputy chairmen of the major State Committees and a few other officials; and the chairmen of the 15 republican Councils of Ministers. It is clearly an unwieldy body, which in fact meets only a few times a year and clearly has primarily a formal role. It has a presidium, made up of its chairman, first deputy chairmen, and deputy chairmen, and it is possible for other individuals to be coopted on an *ad hoc* basis (we have no information on whether such cooptation does in fact takes place). The presidium is an active body, seemingly meeting weekly and carrying on the real business of the state apparatus.

The presidium operates an extensive network of permanent commissions. They are formed on a sectoral basis, with one each for heavy industry, machine building, the military–industrial complex (the VPK, on which more later), chemicals and petrochemicals, light and food industry, construction, agriculture, and transport and communications. Each commission is headed by the relevant deputy chairman of the Council of Ministers, and includes in its membership representatives of the Central Committee apparatus, the top-level 'functional' economic management bodies (Gosplan, GKNT, etc.), and the relevant ministries. It will be noted that there is no commission specifically concerned with science and technology. Each sectoral commission is responsible for technological development in its sector. This means that the Academy of Sciences is not covered in the permanent commission system. The commissions seem to exist primarily as a forum in which economic policies, including technology policies, are worked out and agreed between all interested parties before being passed up to the presidium of the Council of Ministers for formal decision. We have very little information on how they operate, but to the extent that they are the only formal bodies in which representatives of the party and 'functional' and 'line' apparatuses meet, we can hypothesise that they are important.[23]

The presidium has an apparatus, known as the Administration of Affairs (*Upravleniia delami*), with an elaborate organisational structure. Science and technology is represented in this structure by the sector for new technology, one of six sectors of the Administration's subdepartment for central planning organs.[24]

State Committee for Science and Technology

Some mention has already been made of GKNT. It could be described as the USSR's Ministry of Science, and more detailed attention needs to be devoted to it. It was established in October 1965 as part of the industrial reorganisation called for by the September 1965 Central Committee plenum.[25] The new body replaced the State Committee for the Coordination of Scientific Research Work and the numerous predecessors of that body.[26] Its tasks were set out in a Law passed by the Supreme Soviet:

> Article 8. To reorganise the USSR State Committee for the Coordination of Scientific Research Work as the All-Union State Committee of the USSR Council of Ministers for Science and Technology, entrusting it with the preparation together with the USSR Academy of Sciences of suggestions on the basic directions for the development of science and technology in the country, the organisation of the development of inter-branch scientific and technical problems, the provision of scientific and technical information, the control of the application of the achievements of science and technology in the national economy, as well as establishing links with foreign countries on questions of scientific and technical cooperation.[27]

The General Statute for GKNT that appeared in October 1966 repeated these tasks, and included another one: 'to raise the effectiveness of scientific research and to provide for the rapid introduction of the achievements of science and technology into the national economy with the aim of obtaining the greatest economic effect at minimum cost'.[28]

We shall now look in more detail at these tasks. I shall not concern myself here with GKNT's information or foreign relations functions, as important as they undoubtedly are, but shall concentrate on its domestic R&D management functions. They can be summarised under two basic headings: planning and management of implementation.

Planning Soviet science and technology

The planning function has two major aspects. Firstly GKNT has major responsibilities in the area of long-term forecasting and planning, particularly in the applied science area. It is closely involved, with the Academy of Sciences and Gosplan, in drawing up long-term forecasts and national programmes of research. Details are provided in Chapter 7. It does not have a direct management role once such programmes are put into operation, but instead monitors the work of the management body – usually a ministry and its 'head' institute – which it has appointed to head the programme. Many see this as contributing to the weaknesses and problems of the programme approach, primarily a lack of firm central commitment and control, and call for a stronger GKNT role, including the concentration of all programme funds in its hands.[29] Other commentators, including GKNT officials not wanting GKNT to be turned into an operational management body, oppose such proposals.[30]

GKNT then has a more immediate role in determining which areas of R&D will receive priority attention. It is able to allocate high-priority tasks to research institutes, and provide the funding and staff allocations required for their fulfilment. GKNT has the right to introduce changes into the science and technology (S&T) Five-Year Plan during the period of its fulfilment, including allocating additional research tasks. To fund such tasks GKNT has a reserve fund at its disposal. The fund was originally financed from the state budget, making up it is said 2–3 per cent of the total R&D budget.[31] The 1983 CC-CofM decree on scientific–technical progress set up a reserve fund for supplementary funding for all-Union scientific–technical programmes, with the money coming from yearly deductions from the S&T development funds of the ministries.[32] GKNT also maintains a reserve wages fund for the hiring of additional personnel.[33] It is said that GKNT's reserves are an important source of extra funding primarily for Academy institutes.[34] It appears that funds are generally allocated from the reserve after an approach by a researcher rather than on the initiative of GKNT itself.[35]

Managing implementation

We have already seen that GKNT does not have a direct operational role in Soviet S&T. Firstly, it has virtually no research capacity of its own.

Secondly, and more importantly, it is subject to the strictures of Soviet management theory and practice. It is said to be broadly responsible for the state's management of science and technology, but as a 'staff' rather than an operational body. It has no power to give obligatory orders (*direktivnye ukazaniia*)[36] and in practice the situation appears to be that 'line' organisations and agencies can ignore GKNT's decrees and directives with relative impunity.

Despite this there are many ways that it can exercise an important influence over how research is done. Firstly, GKNT is responsible for drawing up the methodological documents that govern the way Soviet science is managed and supervises much of the *naukovedcheskii* (science of science) research that goes into the preparation of decrees and legislation to be issued by higher bodies. For example, it was under GKNT supervision that all the preparatory discussion and drafting was done of the 1983 decree on the reorganisation of research management.[37] It is also responsible for drawing up the methodologies on the evaluation of research work.[38] However, GKNT's powers here might not be as they seem. Many methodological documents are issued jointly with other bodies, and one can be confident that the highly developed Soviet system of bureaucratic consultation ensures that all other interested parties get a chance to have their say before any documents are published.

We have already seen that GKNT can further progress in areas which it has identified as being of particular importance by arranging national programmes of research and directing reserve resources to particular priority projects. As well as encouraging research which it sees as important, GKNT is also in a position to discourage and indeed discontinue research which it considers to have no future. GKNT is empowered to check the work of research institutes and teams and to organise the testing of new products and processes. It conducts *ekspertizy* (tests) of products and processes, usually by organising and supervising inter-branch commissions of specialists from both the developing and user agencies.[39] On the basis of such testing GKNT can recommend further work on the project, its transfer to the production stage or its abandonment. It is empowered to cancel funding allocations to organisations or projects which are not performing adequately, and the money saved is paid into its reserve.[40]

It has always had the right to examine the work of institutes on a broad scale and recommend the closure of those with which it is dissatisfied,[41] although it is a power which it seems to have used only in recent times. In 1981, 'on the orders of the party and government', GKNT, working with the Academy of Sciences and the ministries, undertook an examination of the entire research and design network, with in the initial stages special inter-branch commissions looking at over 100 branch institutes. This resulted in some ministries deciding to close something like 150 research organisations and to reorganise another 60.[42] The examinations have continued afresh with the coming to power of Gorbachev, with, for example, the campaign in Moscow leading to the closure of seven of the first 14 institutes checked there.[43] This power, if used freely, clearly gives GKNT very considerable power over Soviet R&D. However, it should be noted that it operates actively in this field only when there is a clear government policy that such activity is permitted and expected. It is also always specifically stated in the reporting of these closures that the ultimate decisions were made by the relevant ministries, and not by GKNT.

One of the major problems facing Soviet S&T is the coordination of the activities of jealously autarchical industrial bureaucracies, in particular the ministries. One of GKNT's most important functions is seen as the adjudication of disputes between the different organisations involved in the S&T process. It does this primarily through bringing the interested parties together in meetings and inter-branch commissions. At times GKNT will use its own specialists to adjudicate.

As has already been mentioned, what powers GKNT actually possesses in this and other areas is a difficult issue, one which is rooted in the fundamentals of Soviet management theory and practice. But in general it could be said that GKNT faces the difficulty faced by any Soviet 'staff' agency of a mismatch between responsibility and power. It has no power to give direct orders to operational bodies. Consequently the Soviet press is littered with examples of ministries and other agencies simply ignoring GKNT efforts to settle disputes or push particular products, processes, or programmes.[44] It is claimed by one commentator that GKNT now refuses to involve itself in conflicts between ministries. It will deal only with issues that have already been cleared by all agencies involved; otherwise it sends them back for 'reworking'.[45] Some commentators claim, in what is probably an exaggeration, that GKNT's control of Soviet R&D or adoption of any particular project is a mere formality of virtually no practical significance.[46]

On many occasions it is suggested that the problem is not simply that a valiant GKNT is being ignored by narrow-minded ministerial bureaucrats, but that GKNT itself is less than committed to a tough non-'sectional' approach. There is the general problem that GKNT loses nothing if it gives way to the ministries, and gains nothing if it does not.[47] This is compounded by the fact that GKNT, like most Soviet bureaucratic organisations, is structured on a 'line' or 'branch' basis, i.e. its structure is based on departments responsible for particular and individual branches of industry – the Agriculture Department, Chemistry Department, Light and Food Industry Department, etc.[48] There is evidence that the staff of these departments often come from the ministries they are meant to be controlling.[49] This in itself does not mean that they will necessarily take the side of their former employers. However, there is plenty of evidence in the Soviet press that 'sectionalism' in GKNT departments is a serious problem, with inter-branch problems being ignored and control of individual ministries being weakened. An article in *Sotsialisticheskaia industriia* in 1977 described how the head of the Department of Electrophysical and Electrochemical Processing Methods of the Experimental Research Institute for Metalcutting Machine Tools (ENIMS, the head institute of the Ministry of Machine Tool Industry), through his chairmanship of the relevant section of GKNT's Machine-Building Council, could block any R&D proposals or developments that came from outside his department.[50] In another case a commission set up by GKNT to examine work on a rough-terrain vehicle being developed by the Novosibirsk Aviation Factory and the Siberian Academy's Institute of Hydrodynamics, was dominated by representatives of the Ministry of Automobile Industry's head institute, NAMI, in which the existing rough-terrain vehicle had been developed. The commission recommended cessation of the Novosibirsk project, a recommendation which was accepted by GKNT.[51]

Reforming GKNT

Given the difficulty GKNT has in enforcing its will, or even finding the desire to enforce its will, it is not surprising that there has been much dissatisfaction with its performance and many proposals for its improvement. There were signs of dissatisfaction during the Brezhnev era, with his criticism of GKNT's performance at the November 1979 Central Committee plenum being followed in January 1980 by the dismissal of the then chairman Kirillin. At this time various proposals for strengthening the powers of GKNT were receiving considerable publicity. Many argued that GKNT should be given complete control over the funding of major scientific programmes, rather than making the programme's head institute responsible for funding; indeed that GKNT should take over entirely the operational management of programmes.[52] In 1983, following criticism of GKNT in the August CC-CofM decree on scientific–technical progress for not fully using its coordination and *vnedrenie* powers, Marchuk, the then chairman, called for the granting to GKNT of responsibility for new technology plans, which were so blatantly neglected by the ministries.[53] Other more radical proposals spoke of placing a number of institutes directly under GKNT control, with V. P. Rassokhin in particular championing the 'fourth link' proposal, i.e. that a whole network of institutes, based on the applied institutes that were removed from Academy subordination in the early 1960s, be established under GKNT.[54]

With the coming to power of Gorbachev the future of GKNT, as of so many other Soviet agencies, was not immediately clear. Initially Gorbachev and other top leaders spoke in very non-committal terms about GKNT's future.[55] During 1986 two articles appeared in the press suggesting that GKNT's role had been seriously undermined by the new Bureau for Machine Building. The first appeared in March, and was written by Silaev, the chairman of the Bureau. He stated that the head institutes of the machine tool sector would now be entirely responsible for determining the technical level of the sector, not GKNT or Gosstandart, and that they would be controlled by the Technical Council of the Bureau.[56] The second, appearing in August, was by V. P. Beliakov, the head of the scientific–production association 'Kriotekhnika' (Cryotechnology). He stated explicitly that the new Bureau had made GKNT superfluous.[57] This was followed in September by a statement by Gorbachev that GKNT was still not fulfilling its role, and another in November by Ryzhkov that it had 'not found its place'. Between these two announcements GKNT's chairman Marchuk was moved to the Academy presidency, with no replacement being immediately announced.[58] All this led Western commentators to begin to ask whether GKNT would survive.

But in February a new chairman, Tolstykh, was announced. His background in the defence-related electronics industry, while in keeping with a number of other Gorbachev appointments to central coordination agencies, also suggested a new approach and commitment to GKNT.[59] It is possible to speculate on how much Marchuk's transfer to the Academy from GKNT was a promotion or demotion. But there is little doubt that his replacement, that of essentially an academic by a defence-industry manager with no academic pretensions, indicated impatience with GKNT's muddle-headed attitude to control of the ministries and a determination to exert strong central control. In May 1987 the Politburo called for the raising of the role of GKNT within the context of its discussion of *perestroika* and the planning system.[60] This led to a CC-CofM decree 'On raising the role of the State Committee of the USSR for Science and Technology in the management of scientific–technical progress in the country.' The decree makes no major changes to GKNT functions or powers, and indeed summarises its functions in very similar terms to those used in 1966:

> To devote primary attention to determining the priority directions of scientific–technical progress, the solution of major scientific–technical inter-branch problems, the organisation of the development of principally new technologies, techniques, and

> materials, the establishment of control over the scientific–technical level of the branches of the economy corresponding to the highest world standards, and the formation of the network of scientific-research, project-design, and technical organisations.[61]

This statement removes the reference to *vnedrenie* in the 1966 Statute and better reflects reality, which is that Gosplan maintains responsibility for that stage of the innovation process.

Some interesting points in the decree are the criticism of GKNT for its excessive involvement in day-to-day operations (*tekuchka*), particularly in the management of an excessive number of scientific–technical programmes; the granting to GKNT of responsibility for the development of experimental facilities and scientific instrument-making throughout the country; confirmation of its responsibility for the management of MNTKs (on MNTKs, see Chapter 6); allocation to it of responsibility for issuing *goszakazy* in the S&T field (on *goszakazy*, see Chapter 9); and the transfer to it of responsibility for the Soviet Union's international trade in licences (presumably taking over the functions of Litsenzintorg) and for inventions (with the State Committee for Inventions becoming a committee under GKNT). Finally, in keeping with the general current approach to bureaucratic structures, GKNT is directed to pay more attention to the functional and *svodnye* (collating) departments of its apparatus, at the expense of the branch departments. One doubts that any of these changes will radically change the place of GKNT in science management or remove the ambiguities and problems that have always been evident in its operations. Nevertheless they do indicate a willingness to expand the area of operations of the State Committee and certainly a commitment to its continued major role in Soviet science management.

Gosplan

Gosplan, the State Planning Commission, has no direct responsibility for science and technology. However, to the extent that it has primary responsibility for macroeconomic policy-making and control, it cannot be ignored. It has a Department for Scientific–Technical Progress, headed by V.V. Simakov, a candidate of technical sciences. The Department is important enough for Simakov to be a member of Gosplan's collegium.[62] Although the Academy and GKNT have the initial responsibility for drawing up national programmes in the fundamental and applied sciences and supervising the planning process in the Academy and branch networks, all programmes and plans that are to be included in the national Five-Year Plan have to go through the Gosplan Department. Further, Gosplan has direct responsibility for the planning of *vnedrenie*. A new inter-branch commission for accelerating the *vnedrenie* of particularly important discoveries has recently been established under Gosplan auspices, chaired by its first deputy chairman. It includes members from GKNT, the Academy, Minvuz and other S&T-related agencies. The heads of the branch departments of Gosplan are included as members, with their knowledge of the capacities of the ministries being seen as an important weapon in the hands of the commission. It ultimately achieves its ends by including *vnedrenie* tasks in state plans. Its early meetings considered, among other things, the introduction into series production of the industrial cyclotrons of the Siberian Academy's Institute of Nuclear Physics and the 'self-propagating high-temperature synthesis' technology developed in the Institute of Chemical Physics.[63] The commission is described as an important forum for sorting out conflicts between science and production,[64] but it is too early to say whether it will become a major actor in the *vnedrenie* area.

Conclusion

This chapter has indicated that the USSR has a well-developed bureaucratic apparatus for the consideration and implementation of S&T policy. As one would expect, the top-level party bodies are dominant, with the proviso that science is not a matter of such fundamental political and economic significance that bodies such as the Politburo exercise day-to-day control. That leaves room for the state apparatus, from the presidium of the Council of Ministers down, to play a major role. There appear to be good opportunities for scientists to contribute to S&T policy-making, through both formal membership of executive and advisory bodies and informal access to top-level political leaders. These opportunities tend to be limited, however, to senior scientists of high status and prestige.

It appears probable that there is an administrative contradiction in the division of the science sector for purposes of party and top-level state supervision between the 'ideological' institutions, the Academy and *vuzy*, and the 'industrial' institutions of the branch sector. The other, perhaps more important, unresolved administrative difficulty is that of the power of GKNT. Soviet administrative theory and practice have not solved the issue of the relationship between 'staff' and 'line' management bodies, which seems to guarantee a continuing gap between the powers and responsibilities of GKNT. The consequences of this will be seen when we look in more detail at the 'line' bodies in the following chapters.

Chapter three

The Academy of Sciences

The Academy of Sciences is the oldest scientific institution in the Soviet Union, founded in 1725 on the initiative of Peter the Great. For its first hundred years or so its membership was dominated by imported scientists, including such famous names as Euler and the Bernouillis. With its leaders often being political appointees of dubious scientific or organisational qualities its fortunes fluctuated enormously. Despite this it gradually built up its prestige, particularly in mathematics and the geological and geographical sciences. By 1917, while in great rivalry with the major universities, the Academy was one of the leading scientific institutions of the country, although a very small one by today's standards.[1]

From quite early on the Bolsheviks exhibited a desire to work with the Academy, even at the expense of tolerating its predominantly 'bourgeois' members and leaders. While the Academy's losses through emigration, starvation and hypothermia in the first dreadful years of communist rule were considerable, there was relatively little malice involved on the part of the authorities. Like all other institutions in the Soviet Union the Academy was overwhelmed by the cataclysmic political developments of the late 1920s, which led to the First Five-Year Plan and collectivisation. With the imposition of new Bolshevik managers and members the Academy was firmly reminded of its subjection to political authority, although it showed sufficient desire to compromise that it was spared radical reorganisation or destruction.[2] It was perhaps the war against Hitler's Germany which finally integrated the Academy entirely into the Soviet system. The Academy's scientists showed great dedication and perseverance in their defence work, work which continued after the war in the nuclear weapons and ballistic missile programmes. While its members still display occasional signs of feisty independence, it is essentially a fully fledged, loyal, and powerful part of the Soviet political system.

Structure

The Academy is administratively subordinate to the Council of Ministers, although it is not represented on that body. It appears that the chairman of GKNT, a deputy chairman of the Council of Ministers, is responsible in the Council for all science matters, including the Academy of Sciences. However, the Academy is not hierarchically subordinate to GKNT, either formally or practically. The chairman of GKNT does attend the General Meetings of the Academy. It is unusual for any other senior government official to do so (senior official guests usually come from the party apparatus), although at the 1987 Annual General Meeting the chairman of GKNT attended on one day and the chairman of Gosplan on the next.[3]

The Academy is headed by a president, who is elected by a ballot of full members at a General Meeting for a four-year term. A single official candidate is put forward, in recent years usually by the senior Politburo member responsible for ideology, who openly declares that he is representing the Politburo.[4] The current president, elected in October 1986, is G.I. Marchuk. He is an applied mathematician with interest in mathematical modelling in a wide range of areas. He was head of the Computing Centre of the Siberian Academy, chairman of the Siberian Academy, and chairman of GKNT before his election to the presidency. The importance of this last post makes his election as president somewhat unusual.[5]

All the Academy's previous presidents have also been impressive characters in their own right, as scientists or as organisers. They have never been strongly or overtly 'political' actors, and appear usually to have been prepared, within the realities of the time, to represent the interests of the scientific community, including at times in politically sensitive areas. Indeed a number of Western historians have made the point that at times in the Academy's history the political innocence of the president has meant that there had to be another source of politico-administrative power in the Academy. That source has usually been identified as the Chief Scientific Secretary. He could be described as the chief administrative officer of the Academy, and even today, with presidents of greater political reliability and status, the role of the Chief Scientific Secretary is clearly not negligible. He is in charge of the considerable bureaucratic apparatus of the Academy (in 1986 there were over 1,000 staff in the apparatus of the presidium).[6] The present incumbent is I.M. Makarov. Described as a specialist in the fields of automated control processes, robotics, information sciences, decision-making theory, systems analysis, etc., he is in fact a bureaucrat. He has worked previously as deputy head of the Central Committee's Science

Department and deputy Minister for Higher Education. It remains to be seen whether he will be a force for reform in the Academy.

Despite the obvious and crude manipulation in the past of the Academy rules as far as the appointment of a new Chief Scientific Secretary is concerned, in the modern Academy at least there is no evidence that he plays a dominant or highly political role. As Alexander Vucinich points out, by the early 1960s the Academy was able to a considerable extent to regularise the post of Chief Scientific Secretary and establish the dominance of the president over the secretary.[7] The Secretaries take no public position in scientific debates, although they have been active in important debates over the most desirable organisational structure for the Academy, in particular the relationship between the central apparatus and the Academy's Departments. Here, and in other ways, one can detect a certain defensiveness on the part of the 'bureaucrat' Secretary in his dealings with the iconoclastic academicians.

The Academy has about eight vice-presidents, each responsible for a broad field of scientific endeavour. The collegial body which governs the activities of the Academy is the presidium. Its membership includes the president, vice-presidents, Chief Scientific Secretary, the academician–secretaries (see below), heads of the Academy's union republican and regional bodies, plus other leading scientists. Formal administrative decisions of the Academy, such as the appointment of new institute directors, the establishment of new institutes, etc., are issued in the name of the presidium.[8] We have no way of knowing to what extent serious discussion of these issues takes place. Reports are printed of presidium examinations of the work of specific research organisations and progress in particular areas of research. These are presented in the Academy's journal as genuine and sometimes lively scientific discussions.

There are now 18 disciplinary Departments, grouped into four sections. The sections are rarely mentioned in discussions of the Academy's management functions, and the departments seem to be more important in practice. The departments, listed in Table 3.2, are headed by an academician–secretary (*akademik sekretar'*), with a Bureau made up of leading scientists in the Department operating as a collegial organ equivalent to the presidium at Academy level. In the mid-1960s the Bureau of the Department of General Physics and Astronomy met on average twice monthly, while in 1975 the Bureau of the Department of Geological Sciences met 14 times.[9] The academician–secretary is a senior scientist occupying the post on a part-time basis and assisted in his administrative work by a full-time scientific secretary (*uchenyi sekretar'*) and a small bureaucratic apparatus.[10] The academician–secretary also has a number of deputies.

At a General Meeting of the Academy in 1976 it was decided that a deputy academician–secretary should be made responsible for *vnedrenie* in each Department.[11] In 1987 a new position of deputy academician–secretary for organisational work was established. It is a full-time position, open only to doctors of science or corresponding members of the Academy. (Most of those appointed so far are doctors of science. One had previously been the scientific secretary of the Department in question.[12])

The relationship between the Departments and the central presidium, or more accurately its apparatus, has long been one of the most controversial issues in Academy politics, one that has been resurrected in recent times. It is a debate over centralisation versus decentralisation, which is a matter of permanent controversy in all parts of the Soviet system. In the Academy case the issue is complicated since it is seen as being a matter of a bureaucratic apparatus competing for power with the non-bureaucratic scientists who run the Departments. Since it is considered to be very relevant to the 'democratisation' of the Academy, it is not surprising that it has come to the fore in the post-Stalin period of the mid-1950s and under Gorbachev. The 1950s debate led to some greater powers for the Departments, in the new Statutes of 1959 and 1961.[13]

The issue has again become important since the coming to power of Gorbachev, and particularly since the election of Marchuk as president of the Academy. This presumably has something to do with Gorbachev's general democratisation and decentralisation approach. However, the urge towards decentralisation in the Academy is also connected with extreme dissatisfaction with the performance of its central apparatus. These shortcomings were admitted by the previous Chief Scientific Secretary, although like his predecessor in the 1950s he also claimed that the Departments failed to make adequate use of the powers that they already possessed.[14] The issue has been seemingly somewhat controversial, since a commission set up under academician Kudriavtsev to examine ways of reorganising the work of the apparatus was unable to come to any conclusions.[15] However, soon after his election as president Marchuk promised to give greater independence to the Departments, stating that they would receive more rights in the distribution of financial and material resources, the preparation of institutes' plans, the furthering of international links, and the closing down of ineffective research units.[16] The Departments would also gain the right to confiscate resources from poorly performing institutes and redistribute them to other organisations. In addition, they would be responsible for appointments of deputy directors for scientific matters, scientific secretaries and academic councils of institutes, problem-oriented scientific councils, and efficiency review boards. Central control would be maintained over basic funding, supply, and personnel issues, and over links with governmental organisations. A Temporary General Statute for the Departments containing these provisions was accepted by the presidium at the end of 1986, and confirmed (although still as a Temporary Statute) at the March 1987 Annual General Meeting.[17]

It remains to be seen whether the latest changes will have any practical effect on the distribution of power within the Academy. It is necessary to keep in mind that the Departments dispose of very limited administrative resources, and quite probably are incapable of fulfilling a major administrative role without a major increase in bureaucratic staff. If such an increase were to come about they would presumably quickly lose the goodwill of the scientists.

Siberian Department

Special mention must be made of the Academy's Siberian Department. It was founded in 1957 as part of an effort to disperse the Academy's resources more widely and to encourage the development of Siberia and the Far East. Based in Novosibirsk the Siberian Department has roughly the same formal status as a disciplinary department, although in practice it probably has more independent power. It has its own vacancies for Academy membership, although scientists elected to these vacancies also have membership of a disciplinary department. The Siberian Department has dual administrative subordination to the Academy Presidium and the Council of Ministers of the Russian republic. Its financial resources are allocated from the Russian republic's budget, which probably gives it more freedom over their disposal than the disciplinary departments enjoy. Besides Novosibirsk it has institutes in Tomsk, Iakutsk, and Krasnoiarsk. It was in Novosibirsk that the Soviets pioneered the 'science city' approach, with the Siberian Department's main centre, Akademgorodok, on the outskirts of Novosibirsk, being largely self-contained in terms of research, development and educational facilities, as well as social infrastructure, and attracting to it non-Academy research and educational facilities. (The 'science city' approach has since been widely applied in metropolitan Russia, in such centres as Serpukhovo, Dubna, Obninsk, Chernogolovka, and Pushchino outside Moscow.) The Siberian Department applies a considerable proportion of its resources to the development of Siberia and to applied research in general. However, in international terms its reputation is probably greatest for its fundamental research, particularly in mathematics and physics.

With the success of the Siberian Department the Academy has over the years established further regional centres. These are now situated in Bashkiriia, Buriatiia, Dagestan, the Far East, Kazan, Karelia, Komi, Kola, Leningrad, the Urals, and most recently Noginsk and Saratov. They do not have the formal or informal status of the Siberian Department, although two, in the Far East and the Urals, have just been formally reorganised as Departments rather than Science Centres.[18]

Republican Academies

Each republic of the USSR, with the exception of the RSFSR, has its own Academy. Like the Siberian Department the republican Academies have dual subordination to the USSR Academy and the Council of Ministers of their republic, with the funding coming from the latter. They elect their own members, although these do not have the status of the members of the USSR Academy.

There is always some debate about what the range of work done by the republican Academies should be, with a strong tendency for them to try to set up institutes in all major disciplines. Many see this as inadvisable and leading to an excessive spreading of resources. A degree of specialisation in the republics can in fact be discerned, often determined by special features of the republics' economy or geography.[19]

Just as there has been debate about the proper degree of specialisation within the republican Academies, so their proper relationship with the central Academy has also been controversial. The central Academy has usually had the power of coordination, to be exercised through the republican Academy chairmen's ex officio membership of the central Presidium, the yearly meetings of the Council for the Coordination of the Scientific Activity of Republican Academies, and the occasional hearing of reports by the central presidium on the activities of the republican Academies. However, with the latter becoming increasingly self-sufficient and able to rely on the support of local political élites, many have expressed concern about a dangerously inbred science in the republics.[20]

Marchuk has been particularly concerned about the escape of the republican Academies from central control, and since becoming president has started to do something about it. He has announced union-wide coordinating bodies at departmental level, presumably meaning that each central Academy Department will have a body roughly equivalent to the Council for Coordination. He has also declared that reform of the republican Academies must be carried out according to an all-Union programme.[21]

Branch Academies

Four specialised branch Academies exist. They are the Lenin Academy of Agricultural Sciences (VASKhNIL) and the Academies of Medical Sciences, Pedagogical Sciences and Arts. They have functions within their branch roughly paralleling those of the USSR Academy in general science. However, they have no administrative connection with the Academy, being supervised by the relevant ministries.

Functions

The Soviet Academy of Sciences is unlike any scientific institution in the West. It includes three important functions in a single institution, making it an exceptionally important organisation. Firstly, membership of the Academy is a sign of scientific status and merit, in the way that, for example, a Fellowship of the Royal Society is in the UK; secondly, it has an extensive network of research institutes and facilities; thirdly, it is an important science management body, with management responsibilities for all research conducted in the USSR, not just for that in its own institutes.

Membership

Membership of the Academy is not simply recognition of scientific prestige and merit. It also increases the clout of a scientist in science policy-making and to some degree even increases his or her political status. In addition, it brings with it considerable material advantages – large pay increases, chauffeur-driven car, holiday homes, special medical treatment, etc. For all these reasons membership is eagerly sought and the election process is a matter of considerable importance.

At present elections are held once every three years. In the past the regulations have allowed for elections more often, both yearly and every two years, although under Stalin elections became a rare occurrence. Scientists are elected to either full or corresponding membership (it is possible to become a full member without passing through the corresponding stage). Traditionally elections were held either to vacancies created by the death of an existing member or to new positions established by the Council of Ministers. In the most recent election however vacancies were declared when an existing member, whether full or corresponding, was over the age of 75.[22] The existing member retains his place in the Academy, but presumably no vacancy is declared when he does eventually die. The new procedures are part of Gorbachev's 'youth policy'. They led to an exceptionally large number of vacancies in 1987, although as we shall soon see no great reduction in the average age of academicians.

Vacancies are declared in quite narrow fields of scientific research. For example, in the 1984 election the vacancies in the Department of Geology, Geophysics and Geochemistry were broken up into seven fields – tectonics and the structure of the earth's crust, mineral surveying, surveying geophysics, the development of oil and gas deposits, petrography and mineralogy, geochemistry, and seismology and quake-resistant construction.[23] It seems very likely that sometimes vacancies are declared with a particular person in mind – at the 1984 elections R.V. Petrov and A.S. Isaev were elected unopposed in the Departments of Biochemistry and General Biology respectively. The government can itself declare new vacancies without being tied only to replacing academicians who have died.

Once the vacancies have been declared candidates can be nominated by 'scientific institutions, social organisations, and individual scientists or groups of scientists'. In a 1981 addition to the Academy's Statute it was stated that nominations were to be voted on by secret ballot in the collegial bodies of the nominating organisations (i.e. in the academic council of an institute or university, collegium of a ministry, etc.).[24]

Nominations are then considered in General Meetings of the relevant Academy Departments. Candidacies to both full and corresponding membership are voted on in a secret ballot by the full members of the Department present at the meeting, with a two-thirds majority of 'listed' (*spisochny sostav*) members being required for approval of a nomination. ('Listed' members are all members of the Department, excluding those who are abroad on business at the time of the election.)

The list of candidates is then published, and the final election takes place at a General Meeting of the full Academy. Here a secret ballot of full members of the Academy is held, with a simple majority required to secure election.

A few features of these procedures are worthy of comment and further development. Firstly, there is room in the procedures for party influence over the elections. The party, through the Council of Ministers, is able to declare new vacancies in fields chosen by itself, while it is also able, as a 'social organisation', to nominate candidates. One imagines that this would now happen very rarely, since the party has considerable influence within the collegial bodies of nominating organisations.

Of far greater importance is the *de facto* right of the party apparatus to vet all potential candidates. Zhores Medvedev claims very plausibly that the Academy's party committee and the Central Committee Science Department screen all potential candidates.[25] It is highly unlikely that a candidate unacceptable to the party would ever be nominated, much less appear on the ballot paper. It is now admitted in Soviet sources that in the past worthy scientists have failed to be elected to the Academy for political reasons.[26]

However, the fact still remains that there are secret ballots at all stages of the election procedure, and there would seem to be room for academicians to exercise their independence. It is true that one can find the occasional suggestion that Soviet secret ballots are not always entirely secret, while Sakharov has claimed that if necessary voting figures are doctored.[27] Despite all this, there are a number of cases in which the academicians have struggled very hard against the election of some particularly obnoxious political candidates. The best known example is Sergei Trapeznikov, the Stalinist head of the Central Committee Science Department, who got in as a corresponding member only with great difficulty, and never made it to full membership.[28] A nice case concerns G.K. Skriabin. In 1971 it was decided that he would be the next Chief Scientific Secretary of the Academy, but at that stage he lacked the full membership required by the Statute. Nearly all his predecessors had suffered from the same lack, but their elections to full membership had always been organised expeditiously, if irregularly. Oddly, given that on paper at least his scientific credentials appeared to be a vast improvement over those of his predecessors, in his case the academicians were not prepared to be accommodating. He was unable to gain his full membership, so that he had to remain as acting Chief Scientific Secretary until 1977, when the Statute was changed to make it possible for corresponding members to be Chief Scientific Secretary. The academicians, in what seems to be an even more cheeky gesture, then gave him his full membership at the next election.

An interesting feature of the Academy elections is that there are many more candidates than places, a most rare phenomenon in Soviet elections before the recent introduction of multiple candidacies in elections to Soviets. At the last 1987 election 307 nominations were made to 81 vacancies for full members, and 1,787 to 166 corresponding member vacancies. Thus in most cases the elections are keenly contested.

A simple majority of the full members of the Academy is required for election. The balloting procedure apparently allows for more people to obtain the required 50 per cent than there are vacancies. What happens then is that extra vacancies are created after the election. Thus at the last election 83 full members were elected to 81 vacancies, and 172 corresponding members to 166 vacancies. In the previous elections, in 1984, there were even more *post facto* vacancies, with 55 full members elected to 38 vacancies and 112 corresponding members to 89 vacancies. Despite this, many candidates (of no political sensitivity) fail to gain election, and it is usual to have to make a number of attempts.

Given the competition and rewards involved it is not surprising that the elections are accompanied by a considerable degree of lobbying and behind-the-scenes bargaining. A Soviet journal sums up the pre-election process thus:

> Anyone with even the most distant acquaintance [with the elections] well knows what efforts some candidates had to go to to get the required support among the voters. There were coaxings, and deals, and smiles, and banquets, and presents, and services – big and small – and the swapping of votes, and the direct pressure of the Academy and other authorities.[29]

The most recent elections, taking place in conditions of 'democratisation' and *glasnost*, were unusual in the amount of public lobbying before the election, with some newspapers in particular attacking and mortally wounding the candidacies of some scientists. The president of the Belorussian Academy of Sciences A.A. Zhuchenko was blackballed following public criticism of his role in ecological neglect in Belorussia and his lack of a scientific reputation; M.N. Rutkevich, the former director and 'purger' of the Institute of Sociological Research, was savagely criticised by a number of the Soviet Union's most prominent sociologists; and S.A. Fedotov, the director of the Institute of Vulcanology in Kamchatka, was the subject of a damning article on his dictatorial management style in *Moscow News*.[30] This press involvement was something of which the Academy leadership strongly disapproved.[31] Many others, however, feel that in this area, as in many others, the Academy has a long way to go to meet the requirements of Gorbachevian democracy. There have been calls for the participation of the rank-and-file research staff of the Academy in the discussion of candidates and even for allowing doctors of science and corresponding members of the Academy to vote in the elections.[32]

Although the details that appear in the press are extremely sketchy, it is clear that Academy elections are events of considerable political manoeuvring. Over the short term at least the scientists seem to have considerable opportunity to exercise their independence. In summary it could be said that if someone appears on the ballot it can be taken as a sign of party approval; if he or she is elected it can be taken as a sign of approval from the scientific élite.

Some claim that as more and more careerists slip into the Academy the party gets closer to having a solid majority.[33] We shall have to look more closely at the membership of the Academy to come to any firm conclusions on this point. Table 3.1 shows that there certainly is an increasing bias towards those occupying senior management positions among those gaining, in particular, full membership of the Academy. It shows the position held by scientists at the time of their election. The information comes from Soviet biographies published at the time of election. Only a few years can be shown, since biographical data are not always published on new academicians. In many cases we do in fact know the positions of those listed as 'not given', although I have chosen here to present the data exactly as reported by the Soviets. In almost all cases those shown as 'not given' occupy senior posts in defence R&D establishments. (This does not apply, however, to the 1987 data. Here there are simply gaps in our knowledge, and a considerable number of the 'not givens' are in fact in the humanities.) We see an increasing proportion of new full members occupying posts as institute director, *vuz* rector or higher (usually president of a union republic Academy), primarily at the expense of middle-management scientists (laboratory, department, or sector heads in institutes, or heads of chairs in *vuzy*). Thus it is now far closer to the truth to state that an institute directorship is a prerequisite for Academy membership than the reverse. The proportions for corresponding members are far steadier and more in favour of middle-management personnel, although directors still make up a major component. For both levels of membership and throughout the period covered non-management scientists (senior research worker and professor) have made up an insignificant proportion. It is not possible, and would be quite unfair, to attempt to characterise the scientific merit or spirit of independence of academicians simply on the basis of position held. However, it is not hard to accept that institute directors, whether willingly or not, are primarily administrators with little time for direct involvement in research; they are also more subject to political pressure than those lower in the management hierarchy. However, they are also quite important figures in the scientific world, particularly once they have their Academy membership, with enough prestige to be willing sometimes to take a risk. Middle-level managers, although to some extent burdened by administrative duties, are still essentially scientists running their own research projects in a direct way. They are likely to find themselves somewhat less involved in political 'dirty work', but also do not have the opportunity to assert the independence that can sometimes come with senior scientific posts.

Some take as an indicator of the 'takeover' of the Academy the increasing numbers of party members in the membership. Unfortunately party membership is not something which is revealed as a matter of course. Various estimates have been made of party representation in recent years. Zhores Medvedev's 1982 estimate is 60 per cent of full members and over 70 per cent of corresponding members.[34] The party presence is impressive and almost certainly increasing.

Table 3.1 Post held at time of election to the Soviet Academy of Sciences

	Full		*Corresp.*		*Full*		*Corresp.*	
	No.	*%*	*No.*	*%*	*No.*	*%*	*No.*	*%*
	1946				*1977*			
Director or higher	12	27.9			5	33.3	25	32.9
Deputy director	4	9.3			2	13.3	11	14.5
Middle manager	18	41.9			6	40.0	30	39.5
Senior researcher	4	9.3			–	–	6	7.9
Other	4	9.3			–	–	2	2.6
Not given	1	2.3			2	13.3	2	2.6
	1979				*1981*			
Director or higher	13	56.5	26	32.1	19	43.2	29	32.6
Deputy director	–	–	16	19.8	3	6.8	12	13.5
Middle manager	7	30.4	28	34.6	13	29.5	31	34.8
Senior researcher	1	4.3	–	–	1	2.3	6	6.7
Other	–	–	3	3.7	2	4.5	2	2.2
Not given	2	8.7	8	9.9	6	13.6	9	10.1
	1984				*1987*			
Director or higher	30	54.6			42	51.2		
Deputy director	8	14.5			8	9.8		
Middle manager	13	23.6			18	22.0		
Senior researcher	–	–			2	2.4		
Other	–	–			–	–		
Not given	4	7.3			12	14.6		

In this connection it should be pointed out that the lack of party membership of a senior scientist need say nothing about his political attitudes. Until the late 1940s it was unusual for even the most senior of scientists to be party members. This was partly because of their usually dubious class background, partly because of the ambiguous attitudes at the time towards the intelligentsia in general, and partly because it was simply not part of the 'culture' of society of the time.[35] But this did not mean at all that the bulk of senior scientists were anti-communist or even politically neutral. Such scientists as Nikolai Vavilov were proud to describe themselves as 'non-party Bolsheviks'. On the other side of the coin, now that party membership is at very high levels its value as an indicator of political attitudes or values has been diluted. When virtually everyone is a party member we have to find some other explanation for the differences in attitudes and behaviour which make themselves apparent.

There are two other features of the Academy membership which could affect the political stance of that collective body. Firstly, there is the large and probably growing number of defence scientists gaining Academy membership. We saw in the table above the increasing number of newly elected members for whom no institutional affiliation was given in their post-election biographies – this is a pretty reliable indicator of a defence connection. At the 1984 election biographies were not published. However, we have enough information on 13 of the 55 new full members to know that they are defence connected. (It appears, however, that the number of defence scientists gaining election in 1987 was considerably lower.) The defence component of the Academy membership is by now quite substantial, with the Department of Machine-Building in particular being dominated by them. What we do not know is whether the shadowy defence scientists play any role in Academy affairs, and if they do, what role.[36]

At the moment the age of the members of the Academy is a very sensitive issue, with data being released before the last election showing the very serious 'ageing' of the Academy. Thus between 1977 and 1986 the percentage of full members under 50 years of age dropped from 5.8 to 0.8 per cent and of corresponding members from 17.8 to 7.3 per cent. For those over 75 the increases were from 15.3 to 36.6 per cent for full members and 8.1 to 13.7 per cent for corresponding members.[37] It was for this reason that the new procedure, adopted from the GDR and Hungary, was introduced, according to which vacancies are declared for the replacement of academicians who have reached the age of 75 (with the old academicians still retaining their places). This produced a large number of new vacancies at the 1987 election. However, it apparently did not produce any radical reduction in the average age of the academicians. An academician writing soon after the 1987 election stated that the average age of full members was still 69 and of corresponding members 63.[38] In this election two scientists born before 1916 were elected full members, 17 born between 1916 and 1925, 25 between 1926 and 1930, 20 between 1931 and 1935, and six after 1935 (with 11 unknown). While these new members were not young enough to get the average age down, they do show a strengthening of the inevitable generational change in the Academy. Over 70 per cent of these new members were born after 1925. This hardly makes them youthful, but it does mean that in terms of education and adult life they can be described as belonging to the post-war generation, although not yet the post-Stalin generation. It also means that those who are party members joined the party after the war. Their joining is therefore more likely to be for 'careerist' reasons than for reasons of ideological enthusiasm or war-time patriotism.

The takeover of the Academy by a new, if not young, Soviet generation, by science managers and the shadowy defence scientists continues. They are overwhelmingly party members and not obviously part of the traditional scientific élite. How this affects their behaviour once they are through the hallowed portals is hard to say. But there are no signs that the party wants to destroy the scientific prestige of the Academy by stacking it with nonentities or to use it to give a Lysenko-type dominance to particular scientific disciplines. Thus genuine and relatively anonymous scientists, primarily interested in their work at the laboratory bench, are still able to get into the Academy. Further, one assumes that even the most party-loyal science bureaucrats will represent science in the bureaucratic process with energy and commitment, and probably with some success.

Total Academy membership, broken down by Department, is shown in Table 3.2. It is worth noting the dominance of the Academy in terms of numbers by the Physics and Machine-Building Departments. The importance of the Physics Department is hardly surprising – physics has long been regarded as one of the major strengths of Soviet science. The place of the Machine-Building Department deserves more comment. The Department was only recently instituted, as part of Gorbachev's very strong campaign to reorient resources towards the engineering sector of the economy. However, it is essentially the previous Department of Mechanics and Control Processes under a new name. The new Department has had some existing institutes transferred to it from other Departments (the Baikov Institute of Metallurgy, the Institute of Synthetic Polymer Materials and the experimental workshop of the Institute of High-Pressure Physics[39]), and a number of new institutes have been opened under its control. Even without this special attention the old Department of Mechanics had been the largest in the Academy ever since its establishment in 1963. It in turn had replaced the Department of Technical Sciences, abolished in 1963 to the relief of many of the Academy's fundamental researchers. All these variations of essentially the same department are characterised by the highly applied nature of the work of their institutes and the high percentage of defence scientists among their membership. On both counts they are resented by other academicians, with that resentment being most evident in the great debate over fundamental versus applied science in the Academy in the late 1950s and early 1960s.

Table 3.2 Membership of the Soviet Academy of Sciences, by Department, August 1988

Department	*Full*	*Corresponding*
Mathematics	23	37
General Physics	39	72
Nuclear Physics	16	30
Physical–Technical Problems of Power Engineering	15	43
Problems of Machine Building	41	85
Information Sciences	17	36
Geology	23	65
Oceanology	11	26
General and Technical Chemistry	30	47
Physical Chemistry and Inorganic Materials	26	45
Biochemistry	21	48
Physiology	10	20
General Biology	7	30
History	14	32
Philosophy and Law	11	32
Literature and Language	11	29
Economics	15	28

Note: The new Department of World Economics and International Relations has not been included since no elections have been held to the Department yet.

The other noteworthy feature of the table is the extraordinarily low membership of the Biology Department. This Department seems still to be suffering the legacy of the Lysenko era. The Department of Biochemistry was originally set up to provide a route into the Academy for non-Lysenkoists and it seems possible that it still plays that role.

Research network

One of the most unusual features, in Western terms, of the Soviet Academy of Sciences is its very extensive network of research institutes and other scientific facilities. Soviet statistics stopped giving data on the numbers of research organisations in the Soviet Union in 1975. In that year there were 246 such organisations in the central Academy, employing 41,836 researchers. There were another 391 organisations in the republican academies. Since then the number of researchers has increased to 60,518 in 1987, and one source gives the number of Academy institutes in that year as 250.[40] Data on numbers of research institutions in the central and republican academies can be seen in Table 6.1 (p. 97).

Academy research is said to attract about 8 per cent of the total Soviet science budget, although the Academy's contribution to Soviet science is almost certainly greater than this percentage suggests, since its institutes tend to be the leading ones in their fields. Here I shall not go into detail on the structures of the institutes or the nature of their internal operations – that is left to Chapters 6–9. I shall simply describe in general terms the areas in which the Academy network operates.

In historical terms the Academy's research effort has been devoted primarily to all areas of fundamental research and the social sciences. It has found itself to a considerable extent left out of nuclear engineering and computer research. The absence of computer research in particular from its agenda is seen as having had serious consequences for the state of Soviet S&T.[41] But throughout the Soviet period the most controversial issue has been the Academy's proper role in applied research. The pressure on it to undertake such research derives from the inadequacies of the non-Academy research network. It is not a role which is accepted by all members of the Academy, and in the 1950s opposition to its increasingly applied role culminated in the reorganisations of 1961 and 1963, in which many applied institutes were removed from the Academy and the Department of Technical Sciences, the most applied department, was abolished.

Since then, however, Academy institutes have been again taking on an applied role. Since 1977 there has been a marked increase in the attention given to *vnedrenie* in the Academy's Presidium, general meetings, and journal. At a general meeting in March 1977 a number of changes were made to the Statute of the Academy, including the insertion of the technical sciences as being within the Academy's competence in articles 2 and 4.[42] Since then some of the institutes removed in the early 1960s have been returned to the Academy and a considerable number of new applied institutes have been set up, a process that is proceeding with new enthusiasm under Gorbachev.[43]

The Department of Machine-Building now supervises the type of work that was the responsibility of the Department of Technical Sciences (and it is now openly stated by the president of the Academy of Sciences that the abolition of the Department of Technical Sciences was a mistake).[44] There has even been talk recently of the possibility of setting up a Sector of Technical Sciences containing a number of Departments or, in another version, a Sector for the Scientific–Technical Problems of Mechanics, Technology, Engineering, and Power.[45] One could expect such proposals to meet with very determined opposition from those elements of the Academy which opposed the Department of Technical Sciences in the Khrushchev-era debates.

The final indications of the Academy's major applied role are, firstly, the fact that nearly all institutes have to rely to a considerable extent on contract research for their funding, and, secondly, that in recent times considerable resources have been devoted to building up the developmental facilities of Academy institutes, with there apparently now even being a programme to build a series of factories within the Academy.[46]

Although these developments have been reported in the Academy press in positive terms, there are public signs of the concern which one can be certain many academicians feel. The main concern seems to be the amount of resources that basically fundamental institutes have to devote to applied research and the proportion of their total budgets that come from contract research, especially against a background of a shortage of money for pure science facilities and work. Figures on the percentage of Academy funding from contracts vary considerably, but do show strong increases. Nolting states that between 1966 and 1970 the contract share for the central and republican Academies combined went from 8.5 to 12.6 per cent, and had reached 20–22 per cent by the mid-1980s. He sets the percentage for the central Academy in the mid-1970s at about 12 per cent.[47] Other sources put the 1980 figure at 20–30 per cent, and 21 per cent for the Eleventh Five-Year Plan (1981–5).[48]

Since contract funds tend not to be allocated for fundamental research, these percentages of contract work have made it difficult for Academy institutes to maintain the level of fundamental research. It was stated at the beginning of 1986 that 40–50 per cent of Academy work in money terms is devoted to applied R&D.[49] It is a situation with which many academicians are once again clearly unhappy. The difficulties faced are revealed in an interview with the deputy director of the Institute of Nuclear Physics of the Siberian Academy of Sciences. This is an institute which develops and builds commercial synchrotrons with considerable success, and so does have access to contract funds. Yet the institute is concerned with, on the one hand, the difficulty it has keeping the applied research share of its work below 25 per cent, and, on the other hand, the lack of funding from the central authorities to build a dedicated synchrotron radiation source.[50]

The present leadership shows signs of sympathy with the fundamental scientists, and the stress is now on the funding of fundamental research in key sectors. Gorbachev said at the June 1985 Central Committee conference on scientific–technical progress that 'priority must be given to the development of fundamental science, for it is this which acts as a generator of ideas and opens the way into new territories. Here the role of the Academy of Sciences must be raised.'[51] It has been set as a target in the Twelfth Five-Year Plan that the level of contract research in the Academy be reduced from the 21 per cent in the previous plan to 14 per cent.[52] At the same time promises have been made of increased budgetary funding for the Academy. It was given an increase of almost 9 per cent in its 1987 funding, exceeding the increase in budgetary allocations to the science sector as a whole by nearly 3 per cent.[53]

Nevertheless one suspects that the supporters of the fundamental sciences are not yet satisfied. Indeed for pure scientists there are grounds for worry in the type of fundamental research which is receiving priority. Gorbachev went on to say in his speech: 'We must turn Academy institutes sharply in the direction of widening research with a technical orientation, and raise their role and responsibility for establishing the theoretical foundations of fundamentally new forms of technology and technique.' The phrase 'giving priority to fundamental research and strengthening its technical orientation' has become a key and regularly repeated one. The signs of concern are there, with such senior academicians as G.I. Petrov, Vitaly Ginzburg, A.V. Fokin, and Aleksandr Prokhorov stating that truly fundamental research must be protected.[54] Marchuk, in his election speech, declared that the most important issue of the moment was raising the role of the Academy in fundamental science, and quite pointedly stated that with the introduction of *khozraschet* and self-financing into industry the responsibility of the branch R&D sector for *vnedrenie* will be greater.[55] In November 1986 he stated openly that the Academy had been excessively involved in production-related R&D.[56]

In terms of allocation of resources there seems to have been an effort to satisfy both sides. After considerable lobbying the particle physicists have received funding for a number of very large facilities, and at the same time there has been a large expansion of the Academy's engineering science network.[57] Our conclusion must be that at the moment the policy is to try to increase funding and priority for both fundamental and applied science within the Academy. The increases in the Academy's budget might be enough to achieve this, and to cut the reliance on contract research at the same time. But if there is not enough money, it is not clear which side will win. The fundamental scientists have won such battles in the past, but they tend to lose the wars. The fate of proposals to set up a Sector for Technical Sciences could be an indicator of where the balance of power lies.

Research management

The third leg of the Academy triad is its research management role. It obviously must have well-developed management procedures just to keep control of its own institutes, plus all the republican Academies. The latter, as has already been mentioned, are coordinated through the

Council for the Coordination of Research in Republican Academies, while the institutes are accountable to the Bureau of their superior department and occasionally report directly to meetings of the Presidium. As we saw above the relationship between institute, Department and Presidium is always a matter of controversy, and rarely more so than at the moment.

The Academy's management functions go far beyond the boundaries of its own research network. The Academy has long had the right to coordinate all fundamental and social science research done in the country, regardless of the institutional subordination of the organisation doing the research. Then in 1976 at the 26th Party Congress Leonid Brezhnev declared:

> The Party values highly the work of the Academy and will raise its role as the centre of theoretical research and the coordinator of all research work in the country.[58]

This was a new statement of the Academy's role, and one that was immediately taken up by the Academy as important. In 1977 the Academy's Statute was changed to recognise the overarching or 'supra-departmental' (*nadvedomstvennyi*) character of the Academy.[59] However, since 1976 there have been complaints about the lack of definition of the Academy's role, and they have again begun to surface under Gorbachev. N.A. Gordeeva wrote in 1985:

> So far the role of the 'big' Academy as the main coordinator of all scientific work in the country, declared in many party and government documents, has still not received adequate legal expression. The influence of the Academy on the choice of scientific priorities (*napravleniia*) and on the effectiveness of *vuz* and especially branch science remains insignificant.[60]

While there has been no doubting the steady increase in the Academy's role outside the traditional boundaries of fundamental and social sciences, there is still no sign of any precise and enforceable legal power. To a large extent the change has been in the amount of applied research work done by the Academy itself, as outlined above, rather than in major increases in its management role over non-Academy organisations.

Despite this the Academy's influence in this area is not entirely insignificant. One assumes that there is considerable informal contact and consultation between senior members of the Academy and other science management bodies such as GKNT, Gosplan, and individual ministries. In this regard it should not be forgotten that many of the elected members of the Academy work in institutes and management bodies outside the Academy's network. The more formal powers of the

Academy can be listed under five categories: scientific–methodological leadership (*nauchno-metodicheskoe rukovodstvo*) of non-Academy organisations, Scientific Councils, project planning, *vnedrenie* suggestions, and inter-agency organisations.

Scientific–methodological leadership of non-Academy organisations

The Academy has the right of scientific–methodological leadership when and to the extent that non-Academy organisations are involved in fundamental research. This covers some of the most important non-Academy research institutes. In 1986 29 branch institutes were under such Academy leadership, whether of the Presidium itself or a Department.[61] K.V. Frolov, the academician secretary of the Department of Machine-Building and the most enthusiastic advocate of an increased applied role for the Academy, has called for further institutes to be put under the Academy's scientific–methodological leadership and for the whole issue to be seriously reconsidered.[62]

The big question, however, is what does scientific–methodological leadership entail. It does not give the right to give direct operational orders to the subordinate institutes. While the Academy would appear to be guaranteed the right to examine the work of these institutes, both on the spot and in meetings of Academy organs, and on rather rare occasions does so, it can only make recommendations as to future action. It is generally considered that the Academy's influence in this area is as slight as its effort.[63]

Scientific councils

Councils and commissions of varying degrees of formality have always had an important role to play in Soviet science management, including within the Academy of Sciences. In the reorganisation of the Academy in the early 1960s scientific councils (*nauchnye sovety*) received considerably increased attention as major new coordinating bodies.[64]

But as is so often the case in the Soviet Union, the councils were not given any powers to fulfil their new functions. They had no powers to give orders in any field of operation or concern, relying instead on making recommendations to the leading organs of the Academy (with of course the Academy in turn having no power to issue obligatory orders to non-Academy institutions). On top of that they had no apparatus of their own. Presumably they relied on the secretarial and other facilities of the chairman's institute.

It is therefore not surprising that the councils did not live up to expectations. They limited themselves to organising meetings, symposiums, conferences, etc., the publication of research findings, and the issuing of occasional reports on progress made in the field. While they thus played a valuable informational role, they played a very minor role in science planning and management.

Criticism of the councils became particularly strong after the accession of Aleksandrov to the presidency of the Academy, with calls for the reorganisation of the whole council network.[65] This more or less coincided with the time when 'programme-goal' planning became very fashionable, and it was thought that the councils should have a major role in this area, moving somewhat from the coordination of day-to-day research to the setting of long-term strategies and priorities.[66] It is certainly true that the councils do have an important role in the preliminary work in drawing up research programmes and long-term forecasts of scientific–technological progress.

But despite this they still suffer from the usual problems of Soviet coordinating bodies – very extensive, ill-defined, and overlapping responsibilities with no power. The traditional calls for an extension of the councils' powers – in funding allocations, personnel movements, priority determination, and allocation of research tasks – are once again much in evidence in the Soviet press, but again there are no signs of their powers actually being extended. Indeed concern has been expressed about their uncontrolled proliferation.[67] In 1987 the Academy set up a commission under vice-president Fedoseev to examine scientific councils. On the basis of the commission's report the presidium directed that the number of councils be cut. More demanding guidelines on when councils could be set up were established.[68]

It appears quite probable that scientific councils play a useful informational and informal coordinating role in Soviet science. And given that we are talking here of Academy councils with responsibilities that go far beyond the boundaries of Academy research itself, to that extent they provide a means for Academy influence over the Soviet science sector as a whole. However, they cannot be seen as major management bodies, lacking the powers to be so. They are certainly no match for the 'sectional' interests and jealousies of the hierarchically based institutional structures, and probably not even for the institution-based scientific 'schools' that dominate Soviet science.

Project planning

Project planning of R&D has an impressive history in the USSR, particularly in the bomb and space programmes. However, in the guise of 'programme goal planning' (*programmno-tselevoe planirovanie*), it has been adopted on a broad scale only since the late 1960s. I have dealt with it in more detail in another publication and later in this book (see Chapter 7). Here I shall briefly describe the Academy's role in project planning, which takes on two major aspects. Firstly, it takes part in the drawing up of programmes; secondly, it manages some of them.

Long-term economic plans – the one adopted at the 27th Party Congress runs to the year 2000 – are supposed to form the basis for five-year and annual economic plans. One of the most important inputs into the long-term plan is the 'Complex programme for scientific and technical progress for the next twenty years', which is drawn up by a joint Scientific Council of the Academy and GKNT, under the chairmanship of the senior vice-president of the Academy. The detailed work on the 'scientific and technical problems' section of the programme is headed by a GKNT deputy chairman and conducted by 16 special commissions under GKNT. Members of the Academy are of course members of these commissions. The Academy is specifically responsible for the 'socio-economic problems' section of the programme.[69] The Academy also has an important although not leading role in drawing up the 'programmes for the development of new technology', which are included in the Five-Year Plan for Scientific and Technical Development, a component part of the national five-year production plan.

The Academy is entirely responsible for the five-year plans for natural and social science research, containing research for which no practical application is foreseen in the five-year period. The Academy has the right to receive the draft plans for work in the natural and social sciences from the republican Academies and ministries and agencies, as well as from its own institutes. The projects in the five-year plan are implemented by means of coordination plans drawn up by the Departments and their Scientific Councils.[70]

The Academy is responsible for the overall management and control of projects contained in the natural and social sciences plan. However, GKNT exercises overall management of the other programmes. The operational management of a programme is usually in the hands of a nominated 'head institute', and even in the case of GKNT's applied programmes this head institute will sometimes be an Academy institute. Of the 170 S&T programmes of the Eleventh Five-Year Plan an Academy institute was the 'head institute' in nine cases (while Academy institutes were involved in 120 of the programmes).[71]

Clearly if the project planning system were working well it would give the Academy very considerable opportunity to influence the long-term development of Soviet science and technology. However, in recent times there have been signs of increasing dissatisfaction with project planning. But it is usually the branch ministries which are most strongly blamed for subverting the approach, suggesting that the Academy might find the project approach a useful way of obtaining influence over the planning and management of Soviet R&D.

Vnedrenie suggestions

The Academy has and exercises the right to prepare lists of suggestions on which research should be included in economic plans, including production plans. Suggestions are variously described as being given to GKNT, Gosplan, the Council of Ministers, and individual ministries. The Academy complains about the lack of notice taken of its suggestions, although it has also been admitted that because of poor developmental facilities in the Academy the proposals put forward are not always well prepared.[72] In 1981, at the beginning of the Eleventh Five-Year Plan, the Academy sent 188 suggestions to GKNT, Gosplan, and 75 other ministries and agencies. No reply was received in 48 cases; there were 31 refusals; and 109 suggestions were accepted.[73]

Inter-branch science–production units

The final method of involving the Academy in research management, and specifically in encouraging the adoption of its research in production, is the setting up of various forms of joint or inter-branch science–production units, often funded by an industrial ministry with scientific control exercised by the Academy. These include the joint laboratories of the so-called '*vnedrenie* belt' in the Siberian Academy, the 'voluntary' inter-branch science–production associations pioneered by the regional centres of the Ukrainian Academy, and the most recent development, the inter-branch scientific–technical complexes (MNTKs).

The Siberian Academy pioneered joint Academy–industry work with its '*vnedrenie* belt', within which industrial ministries built, equipped and funded developmental facilities in Novosibirsk, with the relevant Siberian Academy institutes providing the scientific input and management. However, it is now generally acknowledged that the 'belt' has not been a success, with the ministries, as is their wont, refusing to give the Academy a free hand in the running of the joint facilities and indeed using them for their own production plan fulfilment purposes. A number of the joint facilities have in recent times been converted into purely branch facilities, under the complete subordination of the ministries.[74] The talk now in the Siberian Academy is of following the Ukrainian example by concentrating on building up its own developmental facilities.[75]

It should be noted, however, that despite this perceived failure of the '*vnedrenie* belt' the dual-subordination system has been adopted in other areas, including the Ukraine. They are also popular in Belorussia and the Urals Science Centre.[76] Indeed in the early 1980s ministries were funding experimental bases in five institutes of the Belorussian

Academy, and in doing so were providing half of the Academy's capital funding for the Eleventh Five-Year Plan.[77]

The first 'voluntary' inter-branch science–production associations and complexes were set up in the Western Science Centre of the Ukrainian Academy in L'vov *oblast'* in the late-1970s. They consist of groups of research institutes, design and experimental production facilities, and industrial enterprises, usually from a number of different ministries and agencies, linked together on a contract basis. They are headed by a leading scientist, recommended by and usually a member of the Bureau of the Western Science Centre. The planning is done on a regional level, while funding is organised by the Ukrainian Academy with the participation of GKNT. The approach attracted a great deal of attention at the beginning of the decade, and similar organisations were set up in the Ukrainian Academy's Science Centres. A similar approach was also adopted by the Leningrad Science Centre.[78] The 'voluntary' associations now receive much less publicity than they did in the past.

Inter-branch scientific–technical associations received universal and more formal recognition with the adoption of the latest form of R&D organisation, the MNTKs, at the highest party levels in December 1985 when they were approved by the Politburo. (A summary of an unpublished decree was published in *Izvestiia* the next month.[79]) The MNTKs are headed by a major research organisation, supported by a number of other institutes, design bureaus, developmental facilities, and production enterprises, probably from a range of different branches and ministries. They are examined in detail in Chapter 6. Here I shall limit myself to a discussion of the Academy's attitude towards the MNTKs.

Eight head organisations of the 20 MNTKs identified as being in existence in April 1987 were from the Academy network. The Academy also has the right to comment on the long-term value of the draft projects of all MNTKs. Thus, although the MNTKs are the management responsibility of GKNT, the Academy does have an input into their operations during the planning process and through its institutes which operate as head institutes.

Despite this, there is evidence, presented in Chapter 6, of dislike of MNTKs within the Academy. It has been claimed that it has been particularly slow in implementing the organisational measures necessary for their establishment, that the drawing up of the five-year plans for the Academy MNTKs has also been slow, and that in general the Academy MNTKs are significantly weaker than those of branch subordination. It appears that the Academy is nervous about the amount of applied research and production management that would be involved in a full-scale commitment to the MNTKs. In a situation in which the industrial sector is notoriously unenthusiastic about R&D, these problems are likely to arise in any inter-branch endeavours in which the Academy finds itself involved.

Conclusion

The Academy would seem to have ample opportunity to involve itself in the planning and management of Soviet science, including outside the confines of its own research. It is true that the formal powers of GKNT are in many areas greater. But if we accept, as not all do, that relations between the Academy and GKNT are amicable,[80] the arrangement could well suit the Academy. Its opportunities for involvement are considerable while its responsibilities are limited. When its responsibilities are increased, particularly in the area of applied research and science–production links, it begins to get nervous.

It is possible to discern clouds over the Academy – shortages of funds, particularly for infrastructural needs, a takeover by party 'hacks' and 'careerists', the problems of generational change, and excessive responsibilities for applied research and management. However, none of these seems serious enough to threaten the Academy's position as the dominant scientific institution in the Soviet Union and one which deservedly occupies an important place on the stage of international science.

Chapter four

The branch research and development network

The Soviet economy is structurally organised on a 'branch' industry basis. By branch we mean that a ministry exists for each industry or sector of the economy. Such ministries exist at all-Union and union-republic level, although here we shall generally ignore the federal nature of the Soviet politico-economic system. There has been a constant tendency throughout Soviet history for the number of ministries to increase, and for them to be in ever more specialised fields. Although this is a phenomenon against which Gorbachev has been struggling, there are still about 50 ministries at all-Union and union-republican level in the industrial sphere alone. There are also a number of State Committees. Most of these are 'functional' agencies, such as GKNT and Gosplan.

Each ministry is responsible for S&T within its own branch, and has a complex network of management and operational organisations for this purpose. Indeed the vast bulk of Soviet R&D resources go to the branch sector, i.e. the research, design, and developmental units (as well as a small amount of in-house factory R&D) operating within the branch ministry system. Since the ministry R&D networks serve directly the S&T needs of the sector of industry managed by the ministry, most of the work they do is in the applied R&D field. However, fundamental research is by no means excluded. Some major Soviet institutes doing theoretical work are operated by ministries and State Committees. Examples include the Kurchatov Institute of Atomic Energy under the State Committee for Atomic Energy, the State Optics Institute under the Ministry of Defence Industry, the All-Union Electrotechnical Institute under the Ministry of Electrotechnical Industry, and the Karpov Physical Chemistry Institute under the Ministry of Chemical Industry.

The branch R&D sector is important in the proportion of the overall Soviet scientific effort that it absorbs and its dominance of the applied industrial research sector in particular. But it is also important because it is seen by many as particularly backward and inefficient. The ministries are generally blamed for all the ills of the Soviet economy, including its S&T ills. Their negative influence is usually seen as deriving from anti-innovatory inclinations based on a commitment to gross output at all costs (*val*) and to high degrees of autarchy and sectional jealousies (*vedomstvennost'*).[1]

Structure of research and development management

In this chapter I describe the personnel and organisations in the ministries most involved in the management of R&D. This will not include at this stage any discussion of the structure and operations of research institutes themselves, except to the extent that head (*golovnye*) institutes have management functions beyond their own boundaries.

Soviet ministries have a mixed line–staff structure typical of most large bureaucratic organisations. The line hierarchy consists of the minister and his deputies supervising enterprises, usually through 'intermediate link' organisations in the central apparatus of the ministry. The staff structure, usually known in Soviet parlance as the 'functional' structure, consists of central administrative units looking after planning, personnel and wages, finance, supplies, etc. The line structure is dominant, meaning that the 'functional' units have no power to give direct orders even in their own areas of specialisation, but have to operate through the line hierarchy.

The minister

The minister is the absolute ruler within the ministry, bearing full responsibility for all aspects of its work. Presumably the degree of attention he gives to S&T matters depends on the importance they are given by central policy-makers and within the ministry itself. It appears to be rare for the minister himself to take direct responsibility for S&T, although he is usually the chairman of the Scientific–Technical Council (STC, see below).

It is also worth noting that it is not uncommon for ministers to have some sort of research or technical background.[2] It would be clearly wrong to see ministers as necessarily technical neophytes, without understanding of or access to expert advice on technical issues. However, whether their expertise is enough to overcome all the pressures on them to give priority, in terms of resources and management attention, to production is a different question.

Deputy ministers

Ministers usually have a first deputy minister and six or seven deputy ministers. Occasionally the first deputy minister will be the chairman of the STC and in charge of the technical administration (TA) and other technically oriented staff organs.[3]

Some ministries have deputy ministers specifically designated at the time of appointment as responsible for technology. In other cases, while no formal reference is made to S&T responsibilities in the reporting of appointments, specific deputy ministers do in practice appear to specialise in S&T affairs. There is presumably always someone at deputy minister or higher level with broad responsibility for S&T affairs and supervision of staff S&T organs. However, because much R&D is administered along line hierarchies rather than by staff agencies, deputy ministers with line responsibilities are also involved in the S&T affairs of their sub-branch.

Collegium

Soviet ministries are organised on a *edinonachalie* (one-man management) basis, and the essentially authoritarian tone of Soviet political life might suggest that there is little room for even quasi-democratic consultation, discussion, and decision-making. However, there is also a strand in Soviet political and administrative decision-making theory that puts strong emphasis on the collective approach. While decisions within a single organisation are always made ultimately by a single individual, the institutionalised access of a wide range of interested parties to the decision-making process guarantees them considerable influence.

The most important of the collective organs within the ministry is the collegium. It is essentially a consultative body, but it does have the right to arrive at decisions, which are put into effect by an order (*prikaz*) of the minister. The minister has the right to ignore collegium recommendations, or to issue orders without collegium approval (with the collegium having the right to appeal such contentious issues to the Council of Ministers). One doubts that such difficulties arise often.[4]

The collegium usually has 13–15 members.[5] The membership is generally made up of the minister, as chairman, his deputies, and four or five heads of the major staff and line administrations of the ministry. Representatives of 'social organisations', in particular the secretary of the party organisation, are usually invited to attend although they do not have formal membership. People from outside the ministry can also be invited to attend, and very occasionally have formal membership.[6] It appears that the head of the TA is usually a member.

The collegium meets weekly to consider all important issues arising in the work of the ministry, including both major policies decided by higher authorities (such as a reorganisation of the administrative structure) and more 'operational' issues (for example, a decision on which enterprise should first produce a new product). Collegia regularly consider S&T issues. Of the 253 issues considered by the Ministry of Heavy Machine Building in 1966 33 (13 per cent) were on S&T issues.[7] The collegia of most ministries met to consider the implications for their branch of the June 1985 Central Committee conference on scientific–technical progress. The collegium of the Ministry of Chemical Industry met twice in the following months, once to consider reorganisation of the work of research organisations and once to consider its S&T personnel policy.[8] The collegium of the Ministry of Instrument Making considered and approved the plan for the transfer of the ministry's research institutes into production and scientific–production associations.[9] As an example of a more 'operational' S&T issue the collegium of the Ministry of Non-Ferrous Metallurgy met to consider what to do about the failure of new aluminium electrolysis technology in the Tashkent Aluminium Factory. The decision was reached to abandon the most adventurous of the new technology but nevertheless not to return to the old machinery. It was also decided not to dismiss the director of the responsible institute.[10]

This report suggests a very concrete and significant role for the collegium. I see no strong reason to reject the picture presented. Some sources have suggested that collegia (and other such consultative bodies) have little influence over decision-making, because of lack of time for discussion and lack of expertise and access to specialised information.[11] However, what we know of the consultation procedures leading up to collegium meetings suggests that there is considerable opportunity for all interested parties to acquaint themselves with documentation and to make their opinions known both before and during the meeting. To the extent that those responsible for S&T (the relevant deputy minister and the TA head) are members of the collegium, they will be plugged into these consultation procedures. It appears that collegium decisions are arrived at by consensus rather than majority voting.[12] Given that in many, and certainly the more important, cases 'consensus' will be what the minister and other important people outside the ministry want, the pre-meeting consultation and negotiation become even more important. Formal membership of the decision-making body presumably helps one establish one's credentials in such preliminary discussions. However, other more general determinants of political and bureaucratic clout also become important, for example the general attitudes of the top party leadership towards S&T at the time.

Scientific–Technical Council

The collegium has broad responsibilities for all aspects of a ministry's work. The STC's responsibilities are confined to the S&T field. We do not have a great deal of information on the STC, but judging from what we do know it is quite an important body, although it is of course only consultative. Its recommendations are put into effect by a decree of the minister, quite probably after discussion in the collegium. Administrative implementation is the responsibility of the ministerial apparatus. The STC is usually chaired by the minister.[13] Its membership consists of 'well-known scientists, highly qualified specialists, production innovators and representatives of the scientific–technical societies and other organisations'.[14] We have no lists of STC members, nor indeed any idea of how large the STC is likely to be. One imagines that the membership includes directors and leading researchers of the ministry's institutes, plus factory directors and chief engineers and some of the branch's more famous 'innovating workers'. It is also clear that ministerial bureaucrats are not excluded.[15]

According to the STC Statute and some commentary they have a surprisingly elaborate structure. The Statute refers to a Bureau, Sections and *ad hoc* commissions. The Bureau is made up of the chairman (presumably the minister), his deputies, individual members and section heads, and a scientific secretary.[16] I have seen no references to such a Bureau in operation. The sections and commissions are headed by leading scientists or the heads of the ministry units in charge of the relevant fields of technology.[17] The STC has a small permanent bureaucratic apparatus (of three in the Ministry of Tractor and Agricultural Machine Building), with the Statute suggesting that it relies on the regular apparatus of the ministry for further logistical support.[18]

The issues considered by the STC can be divided into three broad categories.[19]

Advising on branch science and technology policy

This includes providing forecasts of which fields of R&D are likely to be the most important in the future. In making such forecasts the STC should be particularly careful to take into account foreign developments. The STC is also asked to advise on narrow policy issues. For example, it was on the basis of a recommendation of the STC of the Ministry of Automobile Industry that the minister decided that the ministry would no longer rely on the Ministry of Electrical Industry to develop automobile electronics but would do the work in-house.[20]

Evaluating proposals and checking plans and documentation

It is not clear whether all proposals for new products and processes coming to or arising within a ministry have to be considered in the STC. It is possible that most are. In the Ministry of Power Machine Building the STC examines each year all on-going exploratory (*poiskovoe*) research and STCs usually also examine draft plans of their branches' research institutes.[21] In the Ministry of Food Industry the STC examines all 'methodological and normative-technical documentation'.[22]

Troubleshooting

The STC is expected to use the wide-ranging expertise of its members to provide assistance in S&T projects that are proving difficult. Paul Cocks describes how the STC of the Ministry of Power set up an *ad hoc* commission to consider vibration problems that had arisen in the development of a new turbine. The recommendations of the commission were examined and approved by the STC and then formed the basis of a ministerial decree.[23] The STC seems able to decide the fate of controversial projects. For example, the arrival of a new deputy chairman of the STC of the Ministry of Power led to a complete change in attitude towards and eventual abandonment of a new approach to the construction of earthern dams. (It was not revealed what the basic post of the new deputy chairman was.[24])

It is in some cases of this last type that the 'political' significance of the activities of the STC are most obvious. Technology policy, and in particular decisions on whether individual projects should proceed or not, can be exceedingly contentious in the Soviet Union, and particularly within the branch network. We know too little about the membership of STCs to be able to judge how representative they are of different technological points of view and in particular how independent they are of the technological 'conventional wisdom' ruling within the ministry. The membership of an STC is determined by the minister and his collegium, while it has no formal powers of appeal to higher authorities. There seems little reason to believe that the STCs are truly independent bodies, and one suspects that all too often they are used by ministries to provide authoritative justification for 'monopolistic' behaviour and *vedomstvennost*'.

Technical Administration

The specialist central apparatus agency for S&T management is the TA. (Some ministries use different names for their S&T units, with many in recent times being renamed Chief Technical Administration or Chief Scientific–Technical Administration. I am unsure of the significance of this change, and will continue to use the designation TA throughout this book.)

The TA is essentially a staff agency, in the two senses that its prime responsibility is a functional activity of relevance to all the line activities of the ministry and it has no or very limited direct powers over operational production and research organisations. In most ministries only branch head institutes are directly subordinate to the TA; the rest are subordinate to all-Union industrial associations (VPOs), or whatever the most recent middle-level administrative unit is called (see below). The TAs' influence over non-head research institutions is limited to 'methodological leadership' (*metodicheskoe rukovodstvo*) and 'control'. These are complex and confusing Soviet administrative concepts. While there is rarely much controversy over the range of responsibilities that they entail, the question of what powers they imply is more controversial. In the case of S&T methodological leadership means setting operating procedures and standards in the research process – when and how research projects should be considered completed, the basis on which economic effectiveness should be determined, how bonuses should be allocated, etc. 'Control' is very broad in its scope. Essentially it means 'verifying fulfilment' of research tasks that are legally and properly given to research organisations, including that fulfilment is according to the standards required. The Soviet concept of 'control' has further always included the idea that the controlling agency has the right and obligation to take any measures necessary to ensure that plans are fulfilled, giving potential scope for operational involvement.

But the big question as far as both methodological leadership and control are concerned has always been what powers the staff organisations have to fulfil their broadly defined functions. Some claim that TAs have regulatory powers (*rasporiaditel'nye polnomochiia*) and give obligatory instructions (*obiazatel'nye ukazaniia*), including to line organisations.[25] Other commentators speak in more restrained terms. V.S. Pronina plays down the regulatory powers of functional administrations, claiming that their recommendations and instructions can be put into effect only as orders issued by the minister or his deputy.[26] Another writer tries to have it both ways by claiming that functional administrations do have obligatory powers but that they cannot issue orders (*prikazy*) and that disputes between them and line organisations can only be decided at the highest level.[27]

These are the sorts of ambiguities we should expect in any line–staff system, particularly in the Soviet Union where overlapping responsibilities are a long-standing and conscious feature of administrative practice. The complexities of the TAs' role will be evident even here as we list their main functions.

Responsibilities

The TAs' fundamental responsibility is for coordinating the S&T activities of the branch and ensuring that the branch operates with the most advanced technology possible. Its clearest staff function is the service it provides the ministry's decision-making bodies. If an S&T issue is to be discussed in the collegium, the TA prepares the necessary documentation for the deputy minister responsible for S&T, while one imagines that the TA does the staff and administrative work for the STC. For these purposes, the TA has the right to collect information on S&T issues throughout the branch.[28]

The situation becomes more complicated when we turn to more operationally oriented functions, where the TA finds itself having to work with other, often competing, bureaucratic agencies. Practice seems to vary somewhat over time and from ministry to ministry. Lack of information makes it impossible to chart these differing practices with any degree of confidence, and so an attempt will be made here to generalise.

Planning Overall the evidence suggests that the TA has a strong role in planning the branch's R&D and *vnedrenie* activities. It checks and approves the thematic plans of research institutes and science-production associations (NPOs) (more precisely the research units of the latter).[29] It has special responsibility for fundamental research in the branch (for which it acts as official contractor) and for programme research.[30] It seems that TAs usually have a department for *svodnoe planirovanie* (integrative planning, i.e. for collating numbers of plans submitted by other organisations).[31]

Despite all this a ministry also has a Planning–Economics Administration, one of the most powerful functional agencies in the central apparatus. (In any descriptions we have of the responsibilities of ministers and their deputies, the Planning–Economics Administration is supervised by a more senior official than the TA.[32]) In at least some ministries it appears that input for new technology plans goes directly to the Planning–Economics Administration.[33] Even when, as one suspects is usually the case, the TA compiles S&T-related plans, their collation with production and investment plans (to the extent that this is done at all) takes place in the Planning–Economics Administration.[34]

Planning is always a bargaining process, and in such processes it is important to have enough control over one's own specialised area that an integrated and coherent draft can be presented to higher decision-making authorities. The TA could well be handicapped by its probable relative lack of control over draft plans coming up the line hierarchies. However, once it comes to the plan compilation stage the TA seems to have a sufficient monopoly over S&T planning, an area which has been traditionally kept quite separate from production planning, that one suspects that the opportunities for Planning–Economics Administration interference are limited.

Funding To a considerable extent funding of S&T is subordinate to the planning process. In recent years there has been simultaneously a decentralisation and centralisation of S&T funding in the branch sector.

The decentralisation is in the form of the increasing use of contracts between institutes and enterprises. While such contracts cannot be negotiated and signed by the participating parties entirely free of control from above, supervision appears most likely to take place at VPO level, certainly when non-head institutes are involved. The centralisation is in the form of the 'unified fund for the development of science and technology' (EFRNT) and inter-branch programmes. The EFRNT, and more recent variants of it, is made up of deductions from the general profits of the branch, and funds a large proportion of branch research (see Chapter 7). What limited evidence we have suggests that the TA is responsible for the administration of the EFRNT and other central ministry S&T funds.[35] Inter-branch programmes represent another attempt at the centralisation of branch R&D management and funding. The intention was that a single ministry would be responsible for funding and managing all programme work, no matter how many ministries were involved. This has not happened, with individual ministries funding and managing their own parts of the programmes. However, there has been centralisation to the extent that programme work is managed by the TA and not the line hierarchy.

Ministries have a Financial Administration, which in the S&T context is probably responsible for little more than ensuring that correct procedures are being followed, rather than actually controlling the allocation of funds.

Vnedrenie All sources suggest that the TA bears considerable responsibility for the *vnedrenie* of the branch's R&D. As we have seen above, it is responsible for the compilation of branch-wide *vnedrenie* plans. The TA of the Ministry of Ferrous Metallurgy confirmed the ministry's Basic Statutes on *vnedrenie* procedures, while that of the Kazakh Ministry of Light Industry exercises operational control of the fulfilment of *vnedrenie* plans.[36] However, there are aspects of *vnedrenie* in which the TA could well suffer seriously from the classic Soviet problem of the mismatch of power and responsibility. There is not just the problem of the TA's lack of control over enterprises, but also the problem of its lack of control over the investment process. It has an important role in the planning and funding of S&T but not, it seems, when capital projects (including the rebuilding and renovation of existing enterprises) are involved. Here the Administration of Capital Construction has a dominant role – it allocates investment priorities and resources and supervises the work of project design organisations. It apparently even has the right to veto the inclusion of renovation-based R&D tasks in research plans.[37]

Quality control. The evidence is that the TA has responsibility for quality control in the branch. TAs have departments for standardisation and quality and are described as the head organisations for the development of branch quality control systems, exercising supervision over the branch's central standards institutes.[38] I find it somewhat surprising that the TA is given responsibility for quality control. Quality control is an issue that is of great direct importance to production enterprises, not least because quality ratings determine the prices they can charge their customers. If we see the TA as an innovation-oriented body, its responsibility for quality control could give it considerable influence over anti-innovatory enterprises. That it does not always work this way is evident from the degree of 'fiddling' that goes on with quality ratings, with the TAs being held responsible for such misdeeds.[39] It should not be forgotten that the VPOs also have quality control departments. Their relationship with the TAs' departments is not clear.[40] The same can be said of the effect of the new independent quality control system *Gospriemka* on the TA's role.

Personnel. Personnel has always been considered one of the most sensitive aspects of Soviet management, and one in which the party has an extremely active role. I am not concerned here with the role of the party but with the formally responsible bureaucratic organisations. Ministries have a Personnel Administration and a deputy minister for personnel. These make a powerful partnership, of important political as well as administrative clout. One imagines that their responsibilities cover all areas of the ministry's activities, including S&T. These responsibilities probably include keeping the detailed political and work records which determine the careers of all staff. We have no evidence on whether the TA has any formal or informal role in making S&T appointments. I doubt that it is in a position to exercise a primary role.

It is worth noting, however, that the TA does appear to have responsibility for drawing up and implementing pay and personnel management procedures within the S&T sector. Thus the heads of TAs represented the ministries at recent conferences on the new S&T wages system, despite the existence of a separate Labour and Pay Administration.[41]

Staff and structure

In only one source are we given the number of staff of a TA. Before the reorganisation of the Ministry of Tractor and Agricultural Machine Building in the mid-1970s the TA had 55 of the central apparatus's 604 staff (9.1 per cent), and after the reorganisation 67 of 711 (9.4 per cent). This made it the second largest central administration after the Supply Administration.[42]

The TAs have quite complex internal structures, with their own functional and line departments. The TA of the Ministry of Chemical Industry has a planning department directly under the TA head. There are then three deputy heads, each supervising three departments: (1) the scientific organisational department, processes and apparatus, organic chemical research; (2) technical normatives, environmental protection, inorganic chemical research; (3) patents and licenses, scientific–technical information, polymer materials research.[43]

The heads of the TAs seem to be senior and generally well-qualified personnel. We have already mentioned that they are usually members of the collegium. It is also by no means unknown for TA heads to go on to become deputy or even full ministers. Indeed the career prospects for TA heads could well be better than for other functional administration heads. The majority of ministers and deputy ministers seemingly come from line administration units, having worked their way up from factory management. Most functional administration heads are too specialised in their backgrounds to fit into this career pattern. This is less true of TA heads.

There appears to be no strict rule as to their backgrounds. Many have production backgrounds, and some have research experience. Even among those with a production background this is often in a technical job, such as chief engineer, rather than a pure management post, or in a technologically oriented enterprise. In a number of cases managers with a particular reputation for innovation and progressive thinking have been appointed to head the TA.

Performance

The overall impression is that the TA has extensive responsibilities and considerable powers and bureaucratic clout to back them up. From what we have seen here we would expect it to exercise strong control over branch R&D and to exert an effective influence over the ministry's decision-making and operational performance in the direction of technological development and innovation.

The situation appears to be less rosy in practice. TAs are regularly criticised for failures in the S&T field. They have particular responsibility for adopting and promoting new products and processes and for coordinating branch tasks in inter-branch research programmes, whereas they have less control over capital investment plans and the use to which VPO-subordinated development facilities are put. It is therefore not surprising that most criticisms of TAs are in terms of shortcomings in the former two areas – they are held responsible for the general neglect of new trends in S&T and of developments in broad new technological fields or for opposition to particular products or processes.[44]

Assuming that TAs are not simply being made scapegoats for situations over which they have no control, why is it that they are reluctant to demonstrate greater technological innovativeness? As we have already seen, TA heads usually have a technical background of sorts, and indeed it would appear that TA work is seen as requiring a background in S&T, suggesting a realisation that the TAs have particular bureaucratic values and interests that need to be represented in the decision-making process.

Of course, an R&D background does not guarantee the ability to understand and control in a realistic way all the R&D work being done in a ministry. Nor is an R&D background enough to guarantee a 'progressive' approach to S&T. The problem here is that TA heads, like the great majority of Soviet bureaucrats, spend their careers in a single ministry (with due allowance for reorganisations). The technological positions they adopt tend to be those which have long been accepted in the ministry. We shall soon discuss the conservative technological policies of ENIMS, the head institute of the Ministry of Machine Tool Building, in the machine tool industry. There is no reason to believe that the position of the TA will be any different from that of its head institute. It is certainly noticeable that many of the criticisms of TAs come from other ministries pushing particular products or processes, suggesting that the TAs share their ministries' general technological chauvinism. It is said that one of the reasons why the Ministry of Automobile Industry was unenthusiastic about the use of electronics in cars was that the ministry is dominated by mechanical engineers with no understanding of electronics. A.I. Titov, the head of the Administration for Design and Experimental Work (the ministry's TA), was included among the mechanical engineers.[45] In another case the head of the TA of the Ministry of Electrotechnical Industry, along with other ministry officials, opposed a steel-testing device. One of the device's supporters was described as N. Gorelikov, who while not so identified in the newspaper report is in fact the head of the TA of the rival Ministry for Instrument Making.[46] It seems reasonable to suppose that TA heads give their first loyalty to their ministry rather than to the demands of technological progress. Thus, in the case just cited, opposition to the new steel-testing device from the Ministry of Electrotechnical Industry, including its TA, was based on the fact that the new device would make it harder to bend quality standards, a fundamental 'sectional' interest.[47] In another case the TA was directly implicated in a fight with GKNT over the correct methodology for measuring the economic return of the ministry's R&D effort. Such methodologies are one of the direct responsibilities of TAs, and in this case the implication was clear that the TA was trying to fiddle the figures to make the ministry's performance look better than it really was.[48] In other cases typical Soviet 'family' mechanisms have been used to protect TA heads from outside criticism, strongly suggesting that TA heads are part of the ministry 'establishment'.[49]

Our general conclusion is that the TAs are significant bureaucratic actors within the ministerial apparatus. Their heads are senior and experienced bureaucrats, with special backgrounds appropriate to their functions. However, they have to share their powers with other administrative units whose primary interests are not S&T, while even their own backgrounds more often than not lead to conservative technological positions, fitting easily into the general policies of their ministries.

Head institutes

It is not my intention here to describe the internal operations of head institutes, but only their contribution to branch S&T policy-making and management.

Head institutes are major R&D institutes doing applied research, and often basic research as well, in the areas of work covered by the ministry. Each ministry has a single head institute covering the whole branch, which is usually subordinate to the TA. There are also head institutes for sub-branches; these are usually subordinate to the relevant VPO.

Of serious concern is the power the head institutes have to determine and enforce the technological policy of the branch. They usually work in the major areas of branch R&D, including what fundamental research is done in the branch; they provide the members of the commissions and committees that consider and decide the fate of research proposals and innovations; and they act as lead contractors for major intra- and inter-branch R&D programmes.

The concern arises from their strong tendency towards technological conservatism and narrow-mindedness, protecting their traditional technologies against intrusion from both within and without the ministry. The determination of ENIMS to protect metal-cutting techniques is one of the best-known examples.[50]

All-Union Industrial Associations

The VPOs were until recently the main intermediate link in the line hierarchy between the enterprise, production association and scientific–production association, all at the bottom, and the central apparatus, at the top. Discussion of the 'intermediate link' is made difficult at the moment since there has been a determined campaign to have it abolished. The VPOs (which replaced the chief administrations, or *glavki*, in a similar campaign in the early 1970s) have in fact been abolished in most ministries. In some they have been replaced by *glavki*, and in others by chief production associations (GPOs). In at least one ministry the GPOs have also been abolished, after a remarkably short existence. The aim of administrative reformers is to abolish all such intermediate links in order to reduce bureaucratic control of enterprises and the length of chains of command. The ministries are clearly resisting very hard, through the rapid shuffling of nameplates. One suspects that the intermediate link, under whatever name, will remain in existence. It would certainly be rash to exclude them from a study such as this one, and I will for the moment write as if the VPOs are still in existence.

They are directly subordinate to a deputy minister, with no subordination to the functional or staff administrations. They are each responsible for a single sub-branch (with sometimes some degree of regional specialisation). They usually have direct control over the enterprises, research institutes, and design bureaus of the sub-branch. Their concerns are therefore with both production and R&D. The structure of R&D management within the sub-branch largely mirrors the structure within the branch as a whole, with the VPO's Department of Science and Technology playing the role of the TA (with Planning–Economics, Capital Construction, Finance, and Supply Departments also paralleling central functional administrations).[51]

The Department of Science and Technology is subordinate to the VPO chief engineer,[52] and like the TA it apparently plays an essentially staff role rather than exercising direct operational control over enterprises or institutes. It approves research plans and controls their fulfilment, and draws up recommendations on operating procedures.[53]

There is little evidence of the Department of Science and Technology playing a major or influential role within the VPO. What commentary there is suggests that it plays essentially a 'bookkeeping' role and is in no position to resist the production demands that tend to reign supreme in the VPOs.[54]

It was claimed at the time of their establishment, and consequently, that it was a major function of the VPOs, as line organisations with both R&D and production responsibilities, to bring science and production closer together.[55] However, it appears, and unsurprisingly so, that production concerns have been totally dominant. The general tone and the specifics of the Soviet incentive structure make this almost inevitable, while the bureaucratic position of the VPOs makes it possible for them to put the messages conveyed by the incentive structure into practice. With the recent moves against the VPOs it has now become possible roundly to condemn them as responsible for many of the shortcomings of branch S&T.[56]

Recent changes

I have just mentioned briefly some recent efforts to solve the 'ministry' problem in branch R&D and Soviet economic management in general, primarily through the abolition of the 'middle link'. Internal reorganisation is not seen as the only approach. It is recognised that the ministries and their bureaucrats are not entirely to blame for the shortcomings of technological development in industry. It is admitted that there are problems with the incentive structure, and efforts, even if only of a tinkering kind, have been made over a number of years to increase the incentives of ministries and their enterprises to improve the technological level and quality of production. These efforts have concentrated on success indicators, prices, and bonuses. None of the changes have truly moved the balance of power away from the priority given to fulfilment of physical output plans, and so the results have been lacklustre. Some details are given in Chapter 7.

There are also signs of interest in reconsidering the balance of power between the central ministries and regional authorities, most openly the regional soviets but more realistically the regional party apparatus. However, despite recent decrees on the local soviets, the enormous attention devoted to them at the Nineteenth Party Conference, and some signs of anti-ministerial determination on the part of regional party secretaries, it is still too early to say that they will be given the means to dominate the centralised branch system.[57]

We have already seen uncertainty under Gorbachev as to the future role of GKNT in the supervision of the ministries. After early indications that it could be facing a serious threat to its powers and even continued existence, there now appears to be a top-level commitment to its important and expanded role. However, there has not been enough change in this area to suggest that GKNT is seen at this time as the solution to the problem of ministerial *vedomstvennost'* and opposition to innovation. This becomes particularly noticeable when we consider one new agency that has been recently established.

'Superministries' – the Bureau of Machine Building

A recent addition to the central S&T management bodies are the so-called 'superministries'. These are central administrative bodies supervising much broader sectors of the economy than is usual for the highly specialised Soviet ministries. Calls for such organisations were made by a number of commentators in a number of different forms from the mid-1970s.[58] References to the concept appeared in Gorbachev's early speeches.[59] Action was then taken quite rapidly. The first superministry, and perhaps so far the only genuine one, is the Agroindustrial Commission, in which the ministries involved were actually abolished. Since then a whole range of bureaus and state committees has appeared in all major branches of the economy. Perhaps the most interesting for our purposes is the Bureau of Machine Building. It was announced at an October 1985 Politburo meeting. It covers the non-defence machine-building ministries, with these ministries remaining in existence. Its function is described as essentially a coordinating one, and within that function it has the right to 'issue governmental directives, which are obligatory for ministries and agencies, to examine draft yearly and five-year plans, and according to established procedures redistribute material resources needed for the fulfilment of plan tasks'.[60]

In the limited literature available on the Bureau considerable stress is put on its responsibility for S&T, with the clear major thrust being its control over the S&T policies and practices of the ministries. In an article by the Bureau's chairman, Silaev, it is stated that the head institutes of the machine-tool ministries will deal directly with the Bureau, and that they, not GKNT or Gosstandart, will have complete control of technology policy in the branches. This might sound like an increase in the power of the head institutes. However, it is made clear that they will have to change their ways. Direct reference is made to the need to break the metal-cutting monopoly in the machine-tool industry, suggesting that Silaev believes that putting ENIMS at least to some degree directly under the Bureau will allow the monopoly to be broken.[61] Another enthusiast for the Bureau, the late academician Beliakov, also stated that the Bureau would be able to control the monopolies which are so evident in Soviet R&D and which are likely to become even stronger in conditions of self-financing if not checked.[62] Academician Koshkin, the champion of rotor technology, has also stated that the examination of important new engineering developments by the Bureau will place a control over evaluations arrived at by ministry head institutes. He presumably has in mind among other things the unenthusiastic reaction with which his rotor technology has been met by some ministries.[63]

These tasks seem to require quite a detailed operational role on the part of the Bureau. Thus it organises the competitive evaluation of proposals (*konkursy*), the testing of new developments, and the rejection of those that are found unsatisfactory.[64] But it is clearly not permitted to take over the direct management of operational organisations themselves. Thus a major new Centre for the Application of Microelectronics and Automation in Machine Building being set up in Ul'ianovsk will be under the administrative control of the Ministry of Instrument Making, despite the fact that it is funded by and meant to be used by all the machine-building ministries.[65]

The Bureau has a bureaucratic apparatus of its own, with administrations (*upravleniia*) and STC.[66] However, we have no idea as to the size of its permanent staff, or to what extent its structures are staffed by cooptees from the ministries and institutes. We do know that its chairman, Silaev, and one of its first deputy chairmen, Iu. Konyshev (two first deputy chairmen have been identified), have backgrounds in the aviation industry.[67] This defence industry background of new appointees into senior civilian management jobs has been a feature of the Gorbachev era.

It is clear that the Bureau is intended primarily as an attack on the ministries, particularly in the area of S&T. However, it also has to be seen as some sort of competition for GKNT. Silaev has made it quite clear that he sees the Bureau as taking over some of the functions of GKNT, while Beliakov has specifically stated that in the machine-tool sector the Bureau has made GKNT redundant.

One could be sceptical about the success the Bureau will have, particularly in its control of the ministries. Coordinating bodies in the USSR do not have a happy history when it comes to controlling the ministries, and there is already some evidence of the ministries' ignoring the recommendations of the new Bureau.[68] But perhaps a more important issue in judging the effect of the Bureau on Soviet technological performance is what effect it will have on the 'sectionalism' of the machine-tool sector. By setting up the Bureau the authorities seem to be running the danger of increasing the technological independence and isolation of the machine-tool industry – increasing its susceptibility to 'monopolies' and increasing the gap between the major research institutes (under the Bureau, if it ever in fact manages to establish such control) and production enterprises (under the ministries). Indeed one has to ask whether a specialised and centralised machine-tool sector is a good thing. One can see that a superministry such as the Bureau might be able to reduce 'sectional boundaries' between the ministries that it is controlling. But in the case of machine building the problem is often not the transfer of technology from one machine-building ministry to another but the transfer from the branch machine-building ministry to the branch production ministry. If anything the Bureau of Machine Building will strengthen machine-building 'sectional' interests against the interests of user ministries. The other obvious complaint that could be made about the superministries is that they are simply yet another bureaucratic link in the chain, at a time when bureaucratic control of enterprises and institutes is supposed to be reduced. Not surprisingly complaints have been made about the excessively detailed interference of the Bureau of Machine Building in the work of the branches.[69]

Conclusion

It cannot be said that S&T lacks representation in ministry decision-making bodies or procedures. The STC, TA and head institutes are powerful bureaucratic organisations, guaranteed a say in the deliberations of ministry decision-making bodies. There is no reason to believe that they are unable to defend their own interests within the ministry, while senior S&T personnel also occupy influential positions outside the ministry.

The problem is that these interests tend not to be directed towards up-to-date high-quality R&D. Personal and institutional background determine rather that the interest is in well-established non-disruptive technology that originates within the ministry itself. This, of course, is the sort of technology that the ministry as a whole, including the dominant production and line managers, want. Overcoming such common bureaucratic interests is enormously difficult, and scepticism as to the success of current efforts to do so seems to be justified.

Chapter five

Defence, *vuz* and factory research and development

This chapter can best be summarised as a 'sundries' chapter. It deals with three elements of the Soviet science sector which are somewhat outside the mainstream. Defence R&D can be described in such terms because of the obvious requirements of security and secrecy. *Vuz* and factory research are out of the mainstream, and indeed relatively insignificant, because of peculiarities of the development of the Soviet science sector.

Defence research and development

There are good reasons for treating Soviet military R&D separately from its civilian science – the nature and amount of information available on the military sector and the methodologies required to deal with that information are alone enough to justify such a separation, but other reasons are the very different ways the Soviets handle the two sides of their scientific effort and major differences in quality between the two that might result. These reasons are enough to ensure that there is a reasonably high degree of specialisation among Western analysts with regard to the two sectors. I make no pretence to being on the military side of the specialisation, and consequently this chapter will be very much a summary of work done by others.[1]

One of the dangers of a differentiated treatment of military and civilian R&D is that it might in fact exaggerate the differences between the two. As we shall see there are many similarities, in terms of structures and operations. Further, an important issue is the nexus between military and civilian R&D – how much overlap is there between the two and how much do they depend on each other?

Funding and resource allocation

We shall begin with one of the most difficult aspects of Soviet military R&D – how much money is spent on it. It is a difficult issue not just because of the expected lack of information but also because of methodological difficulties, above all defining what part of the military weapons development and procurement process should be included in the 'science' or 'R&D' category and how various 'dual use' research and technologies should be allocated.

It is generally considered in the West that most military R&D spending comes from within the official 'science expenditure' category reported in open Soviet statistics, with the bulk, although perhaps a declining proportion, coming from the central science budget rather than contract research. It is considered that the amounts contained in the Soviet science expenditure and budget categories are too large to be accounted for by civilian research alone. However, it also seems probable that some military R&D, particularly if we include developmental activities near the production end of the R&D cycle, are funded from non-'science' categories.

In an extended critical summary of available Western estimates of Soviet military R&D spending David Holloway concludes that it is likely to be around 50 per cent of total science expenditure, rather than the rather higher figures some other Western analysts have arrived at.[2] This would make current military R&D funding around 15 billion rubles per annum. Roughly the same percentage is given for the number of scientific and engineering personnel involved in military R&D.[3]

Structure

When looking at military structures we find a major institutional distinction between purely military bodies – running the armies and determining the strategies that direct them – and the defence industry – which researches, designs, and builds their weaponry. The sector as a whole is not surprisingly dominated by the top-level party bodies that control all aspects of Soviet society. The most important is of course the Politburo. Military R&D issues do not feature in the published accounts of Politburo meetings, but it is a very safe assumption that decisions on major military issues, including defence industry issues, are made at that level. The precise responsibilities for defence industry at the top levels of the system are not entirely clear at the moment. Lev Zaikov came to Moscow as a Central Committee secretary for defence industry (from a background in the Leningrad defence industry) but has since taken over the Moscow city party committee. Apparently he continues to have some defence responsibilities. However, the Central Committee secretary who is most publicly active in the field is O.D. Baklanov, with a background in the Ministry of General Machine Building, responsible for missiles and rockets. Zaikov is a full member of the Politburo. Baklanov is not a member. The Central Committee apparatus has a

Department of Defence Industry, which presumably exercises the same sort of control over the defence industry as the other branch Central Committee departments do over their branches (see Chapter 2). It is headed by O.S. Beliakov.

The Ministry of Defence, General Staff and the Armed Services themselves (there are five principal fighting services, the Ground Forces, Air Defence Forces, Air Force, Navy, and Strategic Rocket Forces) are important for military R&D, not because they do much research themselves but because they are the users of the final product of the defence industry. The research facilities that they do have seem to be devoted primarily to determining, in general terms, what that product should be and how it should be used. They do not do the actual technical R&D. Their institutional representatives are the deputy Minister of Defence for armaments and the Central Technical Administrations of the Ministry of Defence (apparently one each for artillery, armour, nuclear weapons, and missiles); the Military Science Administration and Scientific–Technical Committee of the General Staff; and the deputy Commander-in-Chief for armaments of each Service, with his own Technical Administration and Scientific–Technical Committee. All these organisations exist to define the military's weaponry requirements and exercise some sort of control over the weapons as they are developed and produced.

The defence industry proper is organised very much like Soviet industry as a whole. Gosplan is involved in the same way as in Soviet civilian industry. It has a defence industry department and a first deputy chairman with responsibility for military–economic affairs. There is then the Military-Industrial Commission (VKP), which is a commission of the Presidium of the Council of Ministers. It apparently has a very important coordinating role, like the civilian branch commissions of the presidium (see Chapter 2).

The real heart of the defence industry is the nine ministries that specialise in military production:

Defence Industry

ground forces equipment – tanks and other armoured vehicles, artillery, rocket launchers, small arms, optical equipment

Machine Building

conventional ammunition and explosives

Aviation Industry

military and civilian aircraft

Shipbuilding Industry
    military and civilian ships
Electronics Industry
    electronic components and micro-computers
Radio Industry
    radio and radar equipment
Communications Equipment Industry
    radio, telephone, and telegraph equipment, including for space communications
General Machine Building
    strategic missiles, rockets, and space vehicles
Medium Machine Building
    nuclear warheads, uranium mining and processing

Three more ministries, for Chemical Industry, Electrotechnical Industry and Petrochemical Industry, have semi-defence status.

The defence industry ministries are structured in essentially the same way as civilian industrial ministries, as described in the previous chapter. Thus they have a Scientific–Technical Council, Technical Administration, a line administrative hierarchy, and research institutes and production enterprises.

US government estimates put the number of R&D organisations in these ministries at about 450, including approximately 50 design bureaus responsible for major weapons systems and another 250 design organisations working on subsystems and components.[4]

As can be seen from the list of activities above, some of these ministries have important civilian functions, such as the production of civilian ships and aircraft. In fact the list above gives only the barest idea of the amount and range of civilian activity actually undertaken by them. Julian Cooper has done very extensive research into this side of their activities, and shows their very important contribution to the output of Soviet civilian industrial and consumer goods. Thus the Ministry of Aviation Industry produces over one-third of the USSR's vacuum cleaners, and also produces prams and aluminium kitchen utensils. The Ministry of Machine Building is the head ministry for the production of samovars.[5] The ministries are also responsible for the R&D work that goes into this civilian production.

Despite this civilian work the bulk of the effort of these ministries is devoted to defence R&D and production. They run some institutes that carry out basic research that is likely to be of dual application. For example, the State Optics Institute is under the Ministry of Defence Industry and yet it does published and publicised work on holography and refractrometric instruments. It presumably does less publicised work on more sensitive aspects of optics research.

But most of the institutes are entirely unpublicised, the so-called 'post boxes', so known because their addresses are only ever given as post-office boxes rather than a physical location. They do the day-to-day defence R&D, and thereby take a very large slice of Soviet R&D funds and resources. Western analysts guess that they concentrate on applied tasks rather than basic fundamental research.

The probable lack of fundamental research in the closed institutes of the defence industry ministries makes the contribution of civilian institutes and scientists to defence R&D an important issue. By all accounts the contribution is a substantial one. A small indication comes from the Harvard survey of Soviet *émigrés*, a survey which was perhaps likely by the nature of its sample to understate defence R&D involvement. Of its respondents 13 per cent had worked in 'closed' institutes, i.e. purely defence R&D institutes, while 37 per cent had worked on military research. This suggests that 25 per cent worked on military research in open institutes.[6] The 'Farewell Documents', the Soviets' papers on technical espionage illicitly sent to the West, reveal clearly the joint work of defence and civilian ministries on some military projects (see below).

This research could be 'dual use' basic research in Academy institutes, major *vuzy* or branch head institutes. A whole range of basic research in lasers, optics, semiconductors, materials science, etc. is of extreme military sensitivity in the modern world. It seems probable that most of this research is carried on in non-defence institutes, particularly in the Academy.[7] It is probably funded from the Academy's regular budget without any specific contribution from the defence agencies.

Civilian institutes are also liable to take on contract research for military agencies, usually to solve specific short-term problems. According to Alexander it is this aspect of the civilian involvement in military R&D that has increased most dramatically in recent years.[8] The Farewell Documents refer to work done by the Ministry of Non-Ferrous Metallurgy with the Ministry of Machine Building on improving armour-piercing shells, the Ministry of Chemical Machine Building cooperating with the Ministry of Defence Industry on the development of infrared reconnaissance equipment, and the Ministry of Instrument Making working on control instruments for diagnostic equipment for nuclear-powered submarine power plants. Such research is presumably directly funded by the interested military agencies.

Quality of defence research and development

It has long been considered that the quality of the R&D, production, and management in the defence sector is superior to that in the civilian sector. Such a conclusion seems justified given that the Soviet Union is a military superpower in a technological age, despite having an economic base that lags considerably behind those of its rivals.

The most obvious explanation is that the Soviet Union spends far more on defence than its rivals do. The size of the Soviet defence budget is a keenly debated issue,[9] while we have just seen that the R&D share of that budget is one of its murkiest lines. But whatever the actual sums involved it does not seem unreasonable to argue that the Soviets devote a very large proportion of their national resources to defence, and that the general living standards of the people suffer as a result.

But the advantage of the military sector over the civilian sector is not just the simple matter of having more money. Probably more important is that the military has absolute priority when it comes to allocation of supplies, and particularly of high-quality supplies. Shortages and difficulties in obtaining materials are seen by many Soviet scientists as their most serious and frustrating problem. It is one that the military sector faces to a far smaller degree.

Another key factor that many see as contributing to the superior performance of the defence sector, in both R&D and production, is the far greater influence given the user. Defence R&D is far more likely to be based on demands and tasks set by the users – the military agencies – than in the civilian sector. The users remain involved throughout the science–production cycle, this being most obvious through the presence in defence institutes and enterprises of the *voenpredy* (military representatives). These are military officers with special and very strict responsibilities for quality control. A cadre of technologically experienced military officers and officials is therefore built up. In recent times many of the most senior military leaders, for example the late Minister of Defence Ustinov and former Chief of the General Staff Ogarkov, have had career experience as weapons developers and managers of advanced technology programmes. Jerry Hough has pointed out cases in which senior military officers have been moved from operational commands to R&D management posts and back again as priorities have moved between development and deployment.[10]

This practice is an indication of another feature of military R&D that is seen as giving it an advantage – a greater degree of integration between research, development, and production. In fact the integration between R&D and production in institutional terms is not great – R&D organisations and enterprises are usually separate. The important feature seems to be the far greater emphasis on development and design in the defence sphere compared with the civilian sector. In the civilian sector far too great a proportion of R&D spending goes on applied research, with very low levels of funding, resource allocation, and management attention available for development, prototype building, and testing. All too often new products and processes are handed over to production enterprises in an unusable state. This is a problem which seems to have been overcome in the defence sector. There, considerable emphasis is laid on design and development, as can be seen from the fact that many of the best known defence R&D establishments are design bureaus. This applies particularly to the aviation industry. Not only do these bureaus have status and bureaucratic clout, but they also have the facilities to undertake extensive testing of their developments, under the watchful eye of the users.

Another advantageous feature of the defence sector is considered to be the greater willingness there to undertake projects on a competitive basis. Again this is particularly evident in the aviation industry, where different design bureaus and different teams within bureaus put forward proposals for a single task. While we are warned not to exaggerate the degree to which competition does in fact exist in the defence sector, any competition at all puts it ahead of the civilian sector.

All these features are seen as desirable ones by the Soviet authorities, and so over the years they have attempted to transfer them from the defence sector into the civilian economy. Thus the whole 'programme-goal' approach to project planning and management was derived from the more integrated defence approach, as were various other management techniques.[11] In recent years a number of efforts have been made to increase the role of the user in the civilian R&D process, both by a propaganda campaign transferring blame for *vnedrenie* failures from producer to scientist and by specific legislation to give the users a greater role in planning, ordering, and testing new products and processes. The *konkurs* (competition) method of allocating projects is much talked about in recent times and clearly owes its inspiration to defence mythology, if not practice. None of these adoptions from the defence sector of course face up to the major issue of how to apply the funding and supply priority given the defence sector to the civilian sector.

Problems

The assumption so far has been that the Soviet defence R&D sector performs well, maintaining the Soviet Union as a military superpower in difficult circumstances, and certainly easily outperforming the civilian sector. There is a revisionist view, however, that considers that what is little more than an assumption of defence superiority is unwarranted.

Some see features of the Soviet defence R&D system that guarantee Soviet military technological backwardness regardless of the state of the Soviet economy as a whole, and some even consider that in terms of return on investment the defence sector performs worse than the civilian sector.

The most basic point is that military science cannot be completely separated from the civilian sector, from the underlying scientific potential of the country as a whole. If there are shortcomings in the general training of scientists and in the quality and availability of equipment and materials, if there is backwardness in areas of fundamental science and major technologies, such as materials science and computing, those problems must reflect on the defence sector, regardless of the special attention given it.

A further obvious point is that defence R&D is probably as patchy in terms of quality as the civilian sector is. This point is made by Balzer, the most obvious explanation being that there are priorities within the defence sector that ensure that some areas are neglected in terms of funding, resources, and management attention while others get the best of everything and perform accordingly.[12]

But there are also negative features that some see as characteristic of the defence sector as a whole, and some of them are in fact negative aspects of those features which we have just described as advantageous. There are those that consider that the user has too large a role to play in defining the directions of Soviet defence R&D and that the military are too conservative, either through ignorance, distrust of 'ivory tower academics', or inherent caution. It is this conservatism that has brought about the incrementalism and slow response to new developments that some see as negative characteristics of Soviet military technology. (Others, however, see Soviet 'incrementalism' as one of its strengths.) Yet others, including some Soviet commentators, see defence R&D as sharing that conservatism which is so much a feature of the civilian sector (see Chapter 8), and indeed see it as probably deriving from the same problems of 'favouritism' and 'careerism'.[13]

Some analysts point to the difficulties that even the Soviet defence sector has as a result of the institutional separation of R&D and production. We have seen that in the defence sector the neglect of developmental work has been overcome to a considerable extent. There is still a gap between development and production, however, and production enterprises have the same disincentives, albeit to a lesser degree, to undertake innovation and accept new products and processes that is so evident in the civilian sector.[14]

A major problem that many see as inherent to any military R&D sector, but particularly one in a country with the history and traditions of the USSR, is that of secrecy. Clearly strict security is unavoidable in a military R&D system, but equally unavoidable are some negative consequences. Firstly restrictions on life style, travel, and publication are seen as discouraging the best personnel from undertaking careers in the defence sector. These disadvantages are not entirely offset by what are often better working conditions, pay, and research facilities. Harley Balzer's survey revealed a marked preference for civilian research, even amongst 40 per cent of those who had worked in closed institutes. Nevertheless the remaining 60 per cent of that group expressed a strong preference for military research, the most interesting comment suggesting that defence R&D had escaped the morale problems of the civilian sector by giving people interesting and responsible work:

> To work in military industry is more interesting. The assignments are more challenging, more complex, there are more opportunities, better equipment, and, strange to say, more freedom, because there is more responsibility.[15]

We can read in accounts of Western defence laboratories that they are able to build up high levels of morale and community spirit regardless of and sometimes because of their isolation from the scientific community at large.[16] Given what appear to be particularly serious morale problems in Soviet civilian science (see Chapter 9), this could be a major factor in any examination of the quality of Soviet defence R&D.

Some have made the comment, however, that the secrecy involved in military work makes it easier for low-quality personnel to cover up their shortcomings, since their work, and their dissertations for advanced degrees in particular, are not made available for broad peer review.

The restrictions on communications in the defence sector can also mean that defence R&D personnel are unaware of what is going on around them in the general scientific world. One assumes that the defence sector has good access to the world's published scientific literature, but one cannot rely on the published literature to keep up-to-date in modern conditions. Informal communications are required, and these are the sorts of communications that are closed to those working in the defence sector. This breakdown of communications could well be most significant at the boundaries between basic and applied research. In the modern world a lot of what at first glance appears to be rather esoteric basic research can very quickly gain considerable applied defence significance. As we have already seen, basic research is still done essentially in the civilian sector, particularly in the Academy of Sciences. If the defence scientists are cut off from such work, the consequences could be serious.

The Soviets are of course aware of the potential problem and have developed a major scientific–technical information service designed to overcome it. It appears that, while the All-Union Institute for Scientific and Technical Information (VINITI) has essentially a civilian role, the All-Union Institute of Inter-branch Information (VIMI) serves as an information agency for the defence sector and for channelling information between this sector and civilian industry.[17]

Further, one does not get the sense from the Soviet literature, particularly the memoir literature, of an unbridgeable gulf between basic civilian research and the defence sector. In fact one gets the strong impression that senior fundamental scientists have an important role to play in bridging that gap. It is made most clear in Khrushchev's memoirs that senior scientists are expected to play their part in defence work, and that Khrushchev certainly would not trust any scientist, no matter how eminent, who had a questionable attitude towards such work.[18] The Academy of Sciences has usually been dominated by scientists operating on the boundary of basic research and defence R&D, ranging from the 'chief theoretician' of the Soviet space programme Keldysh (president of the Academy, 1961–75) to A.M. Prokhorov, the laser physicist (and until recently head of the Physics Department of the Academy). Many purely defence scientists are members of the Academy, with the Department of Problems of Machine Building in particular being dominated by them (see Chapter 3). It seems probable that many senior academicians have high-level security clearances, that they are members of the various commissions and committees planning and controlling defence R&D, and that they have a watching brief on developments that might be of military significance. We have one example which illustrates one aspect of the sort of role they are expected to play. The engineering academician I.I. Artobolevskii was touring Lithuanian institutes with his protégé K.S. Frolov (now head of the Academy's Department of Problems of Machine Building). They saw some interesting work being done on the piezoelectric tuning of films. Those doing the work complained of the 'usual problems' – 'some didn't understand the work, others thought it not worthwhile' because the precision required would have forced the reorganisation of the relevant production facilities. The two visitors clearly immediately understood the defence optics applications of tunable films and went to a Lithuanian Central Committee secretary who ensured that the work went ahead. It would have been interesting to know which Central Committee secretary, but we are not given that information.[19]

Any discussion of the Soviet defence industry must include something on espionage. We have to ask ourselves how much the Soviets are able to overcome through espionage the basic weaknesses of their economic and technological base and the classification of military and other high-technology work done in the West. This is clearly a particularly murky field. However, the evidence that the Soviets have scored many major espionage coups is impressive. It is a matter of considerable debate as to the role played by espionage in the relative speed with which the Soviets were able to produce the atom bomb.[20] In more recent times we have had a number of COCOM-busting cases.[21] While the details of such cases seem invariably to become controversial, with claim and counter-claim as to the significance of these transfers of technology for the strategic balance, there seems little doubt that some Western businesses have knowingly sold the Soviets embargoed technology. The noisiest case in recent times was Toshiba's sale of technology applicable to the machining of super-quiet submarine propellors.[22]

The most semi-open detail we have on the Soviet technological espionage effort comes from the Farewell Documents, documents handed to French counter-intelligence by an officer of the KGB's T-directorate, the directorate concerned with technological espionage.[23] The documents, dating from the late 1970s and early 1980s, provide considerable detail on the organisation and extent of Soviet technological espionage. They appear to provide much, although by no means all, of the information for the 1985 Pentagon study of the issue.[24] They suggest that the espionage effort is organised by the VPK.[25] It vets, apparently with the close participation of VIMI, 'orders' for documents or pieces of equipment ('samples') placed by the defence industry ministries. The orders are distributed as 'tasks', often with the target organisations in the West named, to five collection agencies: the KGB, GRU (military intelligence), GKNT, the Academy of Sciences, and the State Committee for Foreign Economic Relations (GKES). As the list of acquisition agencies suggests, the methods of obtaining the required items include the suborning of Western citizens by professional spymasters, the large-scale collection of freely available documents and catalogues, and the activities of exchange scientists.[26]

The reports in the Farewell Documents suggest that a high percentage of the material obtained is of little or no value, this applying primarily one assumes to non-classified catalogues, etc. However, the many individual cases of espionage referred to in the documents are striking in that they suggest at times a highly targeted espionage effort, yielding a considerable amount of information and hardware which has been applied directly to specific and important military systems and weapons. Nevertheless the overall effort is tiny compared with total Soviet defence R&D and production. Thus in 1980 392.8 million (convertible) rubles were allocated to 2,980 tasks (not for expenditure in that year, but over the lifetime of the tasks). Actual expenditure in 1979 was 49.7 million rubles. (These figures almost certainly refer to the direct costs of obtaining items, not to overheads.) These sums are very small when compared with a military R&D budget of perhaps 15 billion rubles per annum.

Overall the documents generally support the careful conclusions on the Soviet use of Western military technology reached by Julian Cooper (before the documents became available) that the Soviets do make a conscious and determined effort to obtain Western technology when starting off in a new field. They do this, not surprisingly, with the firm intention of not coming to rely on such technology but in order to give them a flying start and to provide the basis for the development of their own technology.[27] While the Farewell Documents perhaps suggest a more routine, on-going, and narrowly targeted effort than Cooper implies, his general conclusions still seem valid.

One suspects that not even the most highly targeted and successful espionage effort would be enough to overcome all the inadequacies of Soviet technology. Like more open reverse engineering, stealing technology is essentially a conservative procedure, since it guarantees that the thief will continue to lag behind his victim. Further, excessive commitment to a particular approach based on espionage can become disastrous if the espionage source dries up. Finally, no amount of espionage can make up for gaps in technology that make the reproduction of stolen technology impossible. This applies in particular to materials science and computing shortcomings.

Given these problems it is not surprising that – despite a range of Soviet high-technology products, civilian and military, showing suspicious similarities to their Western equivalents – on the whole there are such great differences in the whole philosophy of Soviet military R&D and design compared with the West that a substantial reliance on espionage is unlikely.

Conclusion

It appears that the overall backwardness of Soviet science and technology does reflect on Soviet defence science in a way that ensures that it lags behind the West (which is not to say that the final end-product, the Soviet fighting machine, is necessarily any less effective than Western military forces). In terms of comparisons between Soviet military and civilian R&D there seems little doubt that military R&D is superior. Priority funding and access to supplies must make an enormous difference, given the huge difficulties that ordinary civilian institutes face. The stress on developmental facilities also seems to be a key issue. The involvement of the user means that there is a defined and wanted product towards which the scientists and engineers are working, which combined with greater discipline than in the civilian sector gives a sense of purpose that is enormously important in building and maintaining morale. Many of the problems that are still evident are present to a very considerably lesser degree than in the civilian sector.

Unfortunately for Soviet science the advantages of the defence sector cannot be easily transferred to the civilian sector. Clearly priority cannot be given to everything, and the civilian users still do not have the incentive to take on new products and processes that the military users do. The benefits to be gained from a recognition of the importance of adequate developmental facilities are perhaps transferable, and that lesson does seem to have been learnt. The question of how to build and maintain morale is undoubtedly a key one, and there are perhaps lessons to be gained here from the defence sector, although whether they can be put into practical form is open to doubt.

Research in higher educational institutions

Writing about research in higher educational institutions is particularly difficult at the moment since the effects of major recent reorganisations are not yet clear. The Ministry of Higher Education was recently abolished and a whole string of decrees have been issued over the last year or so. However, one assumes that the fundamental features of *vuz* research will not be changed by these measures, or will be changed in a direction that has been evident for some time. It would appear therefore that some reasonable generalisations can be suggested. Here we shall deal only with the *vuzy*'s contribution to research. Their involvement in educating and training future researchers will be dealt with in Chapter 9.

The most striking feature of *vuz* research is the relatively small proportion of the total research effort it makes up. As the educational network expanded, particularly in the late 1920s and early 1930s, it was considered impossible and undesirable that major resources be committed to research in teaching institutes and universities that were being overwhelmed by students, often with very inadequate preparation for higher education. Many teachers were also too hastily trained to be capable of genuine research efforts.

The essential separation of research and education has continued to this day. There has always been dissatisfaction with this state of affairs. It has been considered a serious waste of resources, particularly of highly trained and qualified personnel. It is also considered that the situation leads to poor teaching, firstly because teachers themselves are not up-to-date with research in their field, and secondly because students gain no experience of actual research practice, at least not on up-to-date equipment.

This has produced an increasing concern to encourage a strong research element in *vuzy* and their teaching programmes. This concern has led to a whole series of decrees, dating back to the 1936 Central Committee decree 'On the activity of, and supervision over, higher educational institutions'. The most important recent decrees have appeared in 1978 and 1987.[28]

Structure

Until recently the chief higher educational management body was the Ministry of Higher Education (*Minvuz*). It was a union-republican ministry, meaning that most of its management work was done by Ministries of Higher Education at the republican levels, with the central Ministry being responsible only for central policy-making and the direct supervision of some of the most important *vuzy*. In March 1988 *Minvuz*, along with the Ministry of Education (responsible for schools) and the State Committee for Professional–Technical Education, were merged to form the State Committee for National Education.[29] It is also a union-republican organisation, and so one assumes that it will operate in essentially the same way as *Minvuz*.

The structure for research in the educational sector is relatively complex. *Minvuz* had a Scientific–Technical Council and a Chief Administration for Scientific-Research Work, and one assumes that that section of the new State Committee responsible for higher education retains the same bodies. The Council is made up of bureaucrats, administrators, and scientists from both within and without the higher educational sector. It is a collegial body charged with providing advice on the research being carried out within universities and teaching institutes. It has a large number of specialised sections and commissions.[30]

The Chief Administration is a bureaucratic unit, set up following the 1978 decree on higher education. It is responsible for the day-to-day bureaucratic administration of *vuz* research.

The Academy of Sciences, potentially at least, has a substantial role in the management of *vuz* research in the fundamental and social sciences, since it is responsible for the planning and coordination of all work in these areas. The presidium of the Academy does in fact meet on occasion to consider the state of research in the higher educational sector and in specific *vuzy*, but one does not get the impression of close administrative or scientific control. This is not to deny that there is a considerable amount of joint work done by Academy and *vuz* researchers or the importance of Academy facilities and personnel in providing impetus to research in particular fields or regions.[31]

Within the *vuzy* the senior responsible official is the deputy rector for scientific research work. The *vuzy* also have advisory Scientific–Technical Councils. All teaching staff are expected to undertake research, using the facilities available within their faculties. However, *vuz* research has become increasingly concentrated in specialised research units and in *vuzy* designated as 'leading' institutions in a series of decrees over the last 20 or so years. Seventy *vuzy* were so designated in the 1978 decree.[32] Their research units range from small laboratories to large institutes. They are staffed both by full-time researchers (about 90,000 in the most recent figures[33]) and teaching staff on fractional or temporary appointments. In 1982 the higher educational sector contained 58 research institutes, over 540 'problem' laboratories working mostly on fundamental research problems, and 770 'branch' laboratories working in specialised applied areas.[34]

Contribution

In 1986 the 894 *vuzy* of the USSR spent 1,586,700 rubles on research, of which 1,356,900 were spent in the dominant *Minvuz* network. (Small amounts of independently administered research are done in the teaching institutes of the State Agroindustrial Commission and the Ministry of Health. In 1974, the last available data, *Minvuz*-subordinate institutions did 87.6 per cent of *vuz* research.) This very modest sum, which probably did not include overheads, represented an average of 4.5 rubles per 'scientific–pedagogical worker' for *Minvuz*.[35] The great bulk of *vuz* research is financed under contract. In 1987 this was the source of 87 per cent of *Minvuz*'s research funds.[36]

It is generally considered that these levels of funding are inadequate and have not allowed the proper development of facilities for research in *vuzy*. The dominance of contract research has led to an overemphasis on short-term minor projects (*melkotem'e*), with a lack of developmental facilities making *vnedrenie* problematical. All recent decrees and Politburo decisions on the matter have demanded substantial increases in the amount of research done in *vuzy*, particularly in fundamental research (up from 13 to 20 per cent of total *vuz* research) and developmental work.[37] To achieve this will require overcoming the traditional shortages of equipment and ancillary staff in the sector. In January 1987 the Politburo promised a fourfold increase in expenditure on scientific equipment in *vuzy*.[38] This, combined with a far greater emphasis on strategic planning,[39] will, it is hoped, produce a far greater and more efficient *vuz* contribution to the Soviet scientific effort.

The Politburo's January 1987 discussion also led to six decrees published in March 1987, each issued jointly by the Central Committee and Council of Ministers. Among other things they were concerned with research done in *vuzy* and the training of students destined for careers in scientific research. We shall deal here with only the first issue. The second is covered in Chapter 9. The decrees called for a doubling or even trebling of the amount of theoretical and exploratory research done in *vuzy* by the end of the current Twelfth Five-Year Plan. It was stated that the appropriate budgetary allocations would be made, while *vuzy* should also allocate their own funds to these purposes, presumably using money obtained through contract research, another area set to grow substantially. It was stated that *vuz* research funding plans, after approval by Gosplan and GKNT, would be included in the yearly state plans. From the beginning of the Thirteenth Five-Year Plan *vuzy* would be allocated resources for the construction and equipping of research facilities under the 'Science and scientific services' head of the state plan. The first change presumably means that the fulfilment of funding plans becomes obligatory, and perhaps reduces the opportunities for *vuz* managements to divert research funds to other purposes. The second change gives the *vuz* research sector access to the science budget.

These changes are clear reactions to perceived problems. They cannot be described as radical, and in some cases we have heard them so many times before that scepticism is more than justified. The changes are clearly not meant to change the nature of Soviet R&D or the *vuz* sector's contribution fundamentally. One suspects that determined implementation will be required if they are to make any difference at all.

Factory research

The factory research sector is a very minor part of the Soviet research effort. In the early 1980s the percentage of scientific workers in enterprises was no more than 3.2. Data published in 1983 show 120,000 people working in factory S&T organisations. These people worked in about 10,000 R&D units in 2,000 enterprises and associations.[40]

Some of these research units are in fact very large fully fledged institutes, based in some of the largest Soviet enterprises. Perhaps the best known is the research institute of Uralmashzavod, the giant engineering factory in Sverdlovsk. In the 1970s this institute contained seven design and four technological departments, a department for automated management systems, an experimental shop, and 49 research laboratories. It employed nearly 6,000 people, including 60 candidates and doctors of science.[41] But most factory R&D units are far smaller and, as we shall see below, have enormous difficulty maintaining and justifying their existence.

Researchers are unhappy about working in the factory sector because it has developed over the years as a sector of low priority and therefore low prestige. Pay scales in the factory sector have traditionally been lower than in independent institutes and *vuzy*, while access to equipment and materials is also worse. In addition, factory researchers complain that they get little time or opportunity to do genuine research as distinct from minor 'firefighting' tasks, the barriers put in the way of completing dissertations being a particular source of complaint. Consequently factory laboratories have enormous difficulty recruiting and keeping quality personnel.[42]

It is not only the scientists who are determined to avoid involvement in factory research. Enterprises are also often more than willing to rid themselves of responsibility for the conduct of R&D, and more especially for making use of it in their production activities. The anti-innovatory attitudes of Soviet enterprises are notorious, and it is hardly surprising that they are happy to have their laboratories hived off as independent institutes.[43]

If they are not hived off they are put to work on minor problem-solving tasks and the settling in of new products and processes.[44] As Ronald Amann points out, in the 1970s not a single factory laboratory in the Ministry of Chemical Industry was engaged in work sufficiently important to be included in the ministry's central coordination plan.[45] The nature of the work being done does not encourage the best scientists to work in the factory sector.

These problems guarantee the factory sector a virtually insignificant role in Soviet scientific research. It is not a situation with which the authorities are happy, and indeed the need to strengthen the factory research sector is one of the constant refrains of *Pravda* editorials, politicians' speeches, etc. The weakness of the factory sector is seen as a major contribution to the *vnedrenie* problems which bedevil Soviet R&D.

Occasionally such concerns take on something of a campaign. Such a campaign has been evident under Gorbachev. He called for the strengthening of the factory sector in his speech to the June 1985 Central Committee conference on scientific–technical progress.[46] Similar demands were made in the Twelfth Five-Year Plan and Ryzhkov's speech at the 27th Party Congress. Ryzhkov announced that there would be a complete reorganisation of branch science in the Twelfth Five-Year Plan, with many more research institutes being moved into the factory sector.[47]

This presumably means that the development of the factory sector will primarily take the form of the transfer of institutes to enterprise subordination within the framework of production associations. It is of course current policy that all branch institutes, other than those classified as 'head' institutes, be transferred into either science–production or production associations. The lessons of the past (summarised in the case of science–production associations in the next chapter) are that the integration of research and production institutions within these associations is of a purely nominal kind, with planning, financial, and administrative independence being jealously protected. The latest policy declarations are not enough to lead one to believe that factory research will become a dramatically more important part of the Soviet science sector in the foreseeable future.

Chapter six

The research institute

Research institutes, in the Academy and branch sectors, are the backbone of Soviet scientific research. The relatively low level of direct involvement of production enterprises and higher educational institutions means that the bulk of research is done by full-time researchers in research institutes.

Establishment

Institutes are usually formed on the basis of an existing laboratory, design bureau or even university faculty. That is, when a new field is opened up or a particular researcher pushes his own specialisation hard, initially a small research unit is set up within an existing institution. If the new unit proves itself or its field becomes of higher priority a fully fledged institute is opened. Quite often institutes are set up around a director, either in recognition of his leading position in a new field of work or as a reward for a politically active and powerful researcher. Institutes founded in such a way have a very 'personalised' nature, with the 'founder' director dominating the institute usually until the day he dies. The consequences will be examined in more detail in Chapter 8.

Numbers

The Soviets have only just begun again to publish data on the numbers of research institutions in their country, after such data disappeared from official publications in 1975. The data published for the end of 1987 showed 5,089 scientific institutions, including *vuzy*. Of these 3,224 represented scientific research institutes and their branches (*filialy*) and geographically independent departments (*otdeleniia*).[1] These data appear not to include Academy institutes as separate organisations, however. They simply list the number of Academies and Academy departments, science centres and branches. Another source states that there are currently 250 institutes in the All-Union Academy of

Sciences.[2] There are a further 1,800 or so design and technological organisations.[3] As can be seen from Table 6.1, the number of scientific institutions has dropped substantially since 1974, but the loss has been among the non-institute categories (in particular independent research laboratories and design bureaus doing research work). Indeed the number of institutes has increased substantially.

Table 6.1 Numbers of scientific institutions in the USSR

	Total (incl. vuzy)	*Institutes (incl. branches and departments)*	*Central Academy*	*Republican Academies*
1950	3,447	1,157		
1960	4,196	1,728		
1965	4,867	2,146	193	349
1970	5,182	2,525	234	394
1971	5,307	2,648	246	396
1972	5,367	2,697		
1973	5,249	2,726	245	391
1974	5,269	2,773	245	391
1980	4,938	3,008		
1985	5,057	3,171		
1986	5,070	3,200		
1987	5,089	3,224	250	

Sources: *Narodnoe khoziaistvo SSSR*, various years; *Vestnik statistiki*, no. 7 (1988), p.64; *VAN* no.10 (1987), p.38

Institutes tend to be concentrated in the major metropolitan centres, particularly Moscow and Leningrad and to a lesser extent the union republican capitals. Thus in early 1987 Moscow had 1,087 research and design institutions, when in 1986 the total number of scientific institutions for the whole of the USSR was 5,057.[4] At the beginning of the 1980s 31 per cent of all USSR Academy institutes were to be found in Moscow.[5] The Soviet residence permit system makes it difficult to decentralise research, since people having residence permits for the metropolitan centres are extremely reluctant to lose them by moving elsewhere. Despite this, substantial R&D potential is found outside the major centres, and strict influx and new construction controls in Moscow and Leningrad in particular since the 1960s make this essential. The Academy of Sciences has long been a major source of decentralisation, with the gradual establishment of independent Academies in all the union republics (Moldavia was the last to get its own Academy, in 1961) and a number of important science centres in such places as the Far East and Siberia. The construction by the Academy of a number of 'science cities' around Moscow and Leningrad (such places as Serpukhovo, Dubna, Gatchina, etc.) in the 1960s and 1970s seemed primarily designed to allow institutes and their staff to be situated close to the cities without breaking influx and new construction regulations.

There are still strong scientific presences in some of the traditional centres of learning, such as Kharkov and Kazan. The major industrial centres, such as Sverdlovsk in the Urals, have also built up strong R&D presences. The final form of decentralisation is the siting of major defence R&D facilities in such provincial centres as Gorky and Saratov, quite possibly for strategic reasons.[6]

Size

Data from 1975 show an average of 229 scientific workers per scientific institution.[7] However, it is not possible to generalise on the size of institutes, since they range from the tiny to the enormous. In fact Soviet scientists and commentators clearly have difficulty agreeing on the ideal size of a research institute. There is periodic criticism of the proliferation of institutes too small to be efficient. These are usually branches (*filialy*) of major central institutes set up in the union republics or other non-metropolitan areas. Thus in 1970 over half the research institutes in Belorussia had less than 200 staff, with most of these being branches of central institutes.[8] It appears that often such branches are opened because the central institute has become too large to manage, and more pressingly because of bans on new employment and construction in the metropolitan areas. In recent years there has been something of a campaign against such branches, and a general move towards the closure and amalgamation of inefficiently sized institutes.[9]

But at the same time that complaints are made about small inefficient institutes many are more concerned about institutes that have grown to unmanageable size. It is claimed that such institutes lead to the breakup of informal scientific 'schools' – headed by prestigious scientists and traditionally the basis of Russian science management – and their replacement by management by 'administrative methods'.[10] Such complaints tend to come mainly from scientists, who call for the breakup of institutes into smaller sizes so that the director can revert to being a research leader rather than an administrator.[11] Others consider that the quality of work in such institutes falls not just because of increasing management difficulties but also because, as staff numbers rise, low-quality workers inevitably come to make up a larger proportion of the total staff.[12] These problems are presumably accentuated by the speed with which many institutes have grown into such giants. The average number of scientific workers per research institution increased from 47.1 in 1950 to 229 in 1975. The figure in 1960 was 84.4, in 1965 it was 141.2, and in 1970 stood at 186.1.[13]

There appear to be a number of reasons for these impressive growth rates. Firstly, they were simply part of the enormous growth in the science sector throughout the world as the 'scientific–technical revolution' took hold. In Soviet conditions of a commitment to centralisation, particularly in the industrial R&D sector, growth in staff numbers tended to be funnelled to a considerable extent into existing institutes.

These were important factors in the external environment. But there were also important pressures within institutes leading to increased staff sizes. There was the simple difficulty involved in getting rid of existing staff – in many cases even after pension age and certainly regardless of performance. Large-scale 'efficiency review' (attestation) campaigns under Gorbachev are only the latest in a series of purges and attempted purges designed to get staff numbers down. They have not been spectacularly successful in the past.

Another aspect of this problem is the difficulty involved in closing down existing laboratories. Such laboratories tend to be personal 'fiefdoms' of the researchers around whom they were established and are closed down only when that researcher retires or, just as likely, dies in harness. This means that administratively and personally the easiest – and perhaps the only – way of moving into new fields is to open new laboratories under new researchers. When added to this is the fact that researchers can only easily get their own projects into plans if they do head a laboratory, it is not surprising that the number of laboratories tends to grow rapidly.

As a rough generalisation it could be said that the best known, most successful, and highest priority institutes tend to be very large, meaning that management is an important aspect of institute life and one with which, as we shall see below, there is considerable dissatisfaction. On the other hand, there are many institutes of less priority which are seen by the Soviets as too small, although one suspects that it is lack of attention which is the problem rather than size *per se*.

Structure

There are a range of legislative and administrative enactments which provide a considerable degree of structural uniformity throughout the system. For example, the Academy of Sciences has a Statute (*Ustav*) for its institutes, a document which is undergoing redrafting at the moment. This uniformity is as yet being only slightly broken down by fashionable attention to non-traditional structures and hierarchies.

The main structural units of the typical institute are, at its top level, its director and his deputies, the scientific secretary, and the Academic Council. Sometimes reference is made to the existence of the 'directorate' (*direktsiia*), a collective grouping of top-level managers. Middle-level structural units usually include departments (*otdely*), sectors, and laboratories. Institutes have a 'functional' or 'staff' apparatus looking after personnel, planning, supply, etc. The institute trade union, komsomol, and party organisations are important 'non-scientific' bodies. Finally, many institutes have some facilities for in-house fabrication of scientific equipment and instruments, with an increasing number having extensive facilities for prototype and even commercial production of their R&D.

Director

The most important person in the institute is the director. The primary management concept in the Soviet Union is *edinonachalie*, usually translated as 'one-man management'. This refers to the total decision-making power and responsibility possessed by a single manager, in this case the director, in any working body. We shall devote considerably more attention to institute directors in Chapter 8, in which we examine their style of management.

Deputy director

Each institute has a number of deputy directors, each with very different functions. The most important in the scientific sense is the deputy director for science (*zamestitel' direktora po nauke*). He is someone of a scientific background,[14] and truly deputises for the director in the management of the institute's research. In cases, not unusual in the Academy in particular, when the director has other higher-level administrative positions the deputy director could well be the *de facto* director.[15] Some of the largest institutes have more than one deputy director for science.

The other deputy directors have non-scientific administrative responsibilities, and they appear to have no scientific background. In fact, while they are called *zamestitel' direktora* in the same way as the deputy for science, in these cases a better translation would seem to be 'assistant' rather than deputy, since they are clearly in no way second-in-command. They are sometimes described as responsible for 'general affairs' (*po obshchim voprosam*), or sometimes for 'administrative affairs' (*po khoziaistvennym delam* or *po khoziaistvennoi chasti*). (Some institutes also have a deputy director for personnel.[16]) As Kaplan makes clear in his article, published in 1961 but seemingly still accurate, such deputy directors are not seen as part of the scientific or even executive hierarchy, and it is not considered essential or even desirable that they have a scientific background.[17] In a sample of 44 Kievan research institutes the deputy director for science was included in the 'directorate' in all cases, and the deputy director 'for general questions' was included in 29 cases.[18] Such deputies are responsible for finance, supplies, apparatus, equipment, furnishings, repairs and maintenance, and other service activities.

Institutes of a highly applied nature or with extensive prototype or developmental facilities have a chief engineer. Depending on the extent of engineering work done in the institute he could well have status equivalent to the deputy director for science.

Scientific secretary

The scientific secretary (*uchenyi sekretar'*) occupies an administrative post, but one which is more directly concerned with the scientific affairs of the institute than the deputy director for general or administrative affairs. He is the secretary of the academic council (see immediately below) and signs the minutes of its meetings. He is a voting member of the council, and is ranked among the institute's top management.[19] He is also in charge of the institute's chancelry, and therefore maintains its files and often its archives.[20] One source states that it is his task to improve administrative procedures (*deloproizvodstvo*) in the institute.[21] It appears that he is also responsible for personnel matters, particularly where it is a matter of staff concern over pay, promotion, etc. The scientific secretary seems to have an important role in controlling the preparation of dissertations and organising their defence. This is presumably a matter of ensuring that the necessary formal requirements have been met, rather than playing the role of a supervisor.[22] One source shows this as a joint responsibility with the director.[23]

Scientific secretaries usually, if not always, have some sort of scientific background. Among a sample of 62 scientific secretaries of institutes in the Siberian Academy 21.0 per cent had no advanced degree, 77.4 per cent had a candidate's degree, and 1.6 per cent a doctorate.[24] As we shall see below the Soviet tradition is that research managers at least maintain the appearance of continuing research work, and scientific secretaries like to be described as active researchers.[25] However, there is some evidence that those accepting such a post are seen as moving out of scientific work. When M.P. Matveeva was called to the office of I.P. Bardin, the director of the Institute of Metallurgy, to be told that she would be the next scientific secretary of the institute (she was not given any choice in the matter), her reaction was recorded as follows:

> To tell the truth, I was not so much frightened as upset: if they are sending me to work as the scientific secretary it means that they rate me as a poor researcher.[26]

The scientific secretaryship is a post of some political sensitivity since the scientific secretary is responsible for implementing politically motivated administrative measures. His control over the presentation of dissertations is particularly important here. D.V. Lebedev, the scientific secretary of the Botanical Institute in Leningrad, found himself in trouble with the local party authorities for accepting the doctoral dissertation of Raisa Berg, the anti-Lysenkoist biologist, into his institute for defence.[27] It seems possible that institute directors, wanting to sit on the fence as much as possible where delicate issues are concerned, will palm responsibility off onto the scientific secretary. Thus Khalatnikov, the director of the Institute of Theoretical Physics, is considered a decent but politically timid person. He insisted that all communications the refusenik Mark Azbel wanted to address to him must go through the scientific secretary, Lev Gor'kov. Gor'kov is presented throughout Azbel's memoirs as a hatchet man.[28]

Academic Council

The Academic Council (*uchenyi sovet*) is the highest collegial body in the institute. It is a consultative (*soveshchatel'nyi*) body, meaning that its decisions can only be put into effect by a decision of a responsible official, usually the institute director. In the past the Council was appointed by the administrative body standing above the institute – the presidium of the Academy or the ministry. There are no formal ex officio members, but membership usually includes the director and his deputy for science, the scientific secretary and the party secretary. It is not unusual for the institute trade union and komsomol heads to sit on the Council also, as well as senior administrators such as the deputy director for administrative affairs and the chief accountant. The inclusion of the latter officials is not always appreciated by the scientists in the institute.[29] A number of other researchers in the institute, usually heads of sectors and laboratories, will then make up the numbers. The total membership is usually somewhere in the mid-twenties. It is noteworthy that care is taken to ensure that a majority are party members.[30] Decisions of the Council are decided in a secret ballot, on the basis of a 50 per cent majority.

The use made of the Academic Council presumably depends heavily on the approach of the institute director. One would expect it to be consulted on the full range of scientific issues facing the institute, ranging from allocation of funds to particular projects to reorganising the institute's structure.[31] Traditionally it had a major role in examining dissertations for advanced degrees, but that function has been largely taken over by special institute commissions and the All-Union Attestation Commission (VAK). The Council still has an important function in examining the progress being made in the institute's projects. It is part of the Soviet tradition that researchers present their work in a rather formal way before meetings of their colleagues, and the Academic Council is one of the forums for such presentations. Some institutes require that every piece of work prepared for publication be presented before the Council before being submitted to a journal or publisher.[32]

But probably the most important current function of the Academic Council is in the area of personnel appointments and reviews. The Council is responsible for decisions on appointments to positions that have been open to competitive recruitment (*konkurs*), on renewals of fixed-term appointments, and on the fate of those who have gone through efficiency review procedures (in the last case acting on the recommendations of special review commissions). One source suggests that Councils spend 60–80 per cent of their time on such issues.[33] It can be a sensitive area, with Councils sometimes being expected to bring to a climax politically inspired campaigns against unwanted researchers.[34]

It is the practical subordination of the Academic Council to the director, regardless of the formal relationship between the two, which is at the basis of the current argument within the Academy of Sciences over work-place democracy. There are those who claim that the best way of democratising Academy procedures is to extend the powers of the Academic Council. This opinion is often a defence against those demanding that considerable powers, including the right to elect institute directors, be given to meetings of all institute staff. It therefore tends to be adopted by representatives of the conservative position. The more radical reformers suggest that such an approach is meaningless, since the Academic Council is effectively appointed by the director and totally under his control.[35] The result appears to have been a compromise, with the Councils being opened up to election by meetings of staff in a secret ballot. This could represent a significant change in the way institutes are managed, although there are some ominous signs that the old ways will not be greatly changed. Firstly, Fedoseev, the chief ideologist of the Academy, in discussing the proposed changes suggested that elections would not be conducted in cases of those holding ex officio membership of the Council. There appears not in fact to be any such thing as legal ex officio membership, but presumably Fedoseev has in mind the key management positions.[36] At the moment such positions do not make up a majority of the membership, and so there could still be room for an anti-management majority. However, as another ominous sign, an institute party secretary, in accepting the inevitability of elections, makes his displeasure very clear and suggests that the new approach will greatly increase the work of the party organisation.[37] Presumably he means that the 'organisational work' required to ensure the elections turn up the right results will become greater and more complex. The possible meaning and implications of these changes will be further discussed later.

Structural subunits

Most institutes are divided in descending hierarchical order into departments (*otdely*), sectors, and laboratories. Sectors and laboratories are described as research units, whereas the department is an administrative unit, presumably meaning that it represents the lowest level of formal aggregation for planning, calculation and payment of wages, work evaluation, etc. Because departments are administrative units, it is said that scientists do not like to be appointed to head them, preferring to head a laboratory or sector, a position which gives them access to the planning and resource allocation process without responsibility for administrative bookkeeping. In practice, however, department headships are often held concurrently by laboratory heads.[38] It is said that laboratory heads have direct access to the institute director or deputy, and that they are almost certainly more important individuals than department heads.[39]

The number of structural subunits in an institute clearly depends above all on the overall size of the institute, but also on the nature and thematic divisions of the institute's research. A survey of 44 Kievan research institutes in the mid-1970s (of which 35 were in the natural sciences and 9 in the social sciences) revealed 1,452 structural subunits, of which 791 were departments. Of these 577 were directly involved in research or research-connected ancillary work. The rest were 'functional' or 'staff' units (on which more in the next section). In the natural science institutes departments ranged in size from three to 157 staff members (in the social sciences from five to 45). Within single institutes departments ranged in size from 12 to 157, four to 56, 12 to 93, etc.[40] In this survey attention then turned to laboratories. The number of laboratories in each institute varied from 13 to 36, with the number of personnel in them ranging from one to 59. There is widespread condemnation of the proliferation of lower-level research units, particularly when such proliferation is accompanied by great inflexibility in closing down or reorienting the work of existing units. They tend to be established in very narrow fields, often around the particular research project of a single researcher. Soviet hierarchical and planning traditions make it difficult for a researcher to undertake official work on his own project without being formally allocated a laboratory for that purpose. The result is that laboratory heads treat their laboratories as feudal fiefdoms, to use the medieval analogy favoured by academician Lev Artsimovich.[41] This has increased the management problems in large institutes, while the proliferation of subunits is seen as a root cause of the problem of *melkotem'e*, i.e. the excessive number of minor research projects being pursued at any one time.[42] These perceptions of uncontrolled proliferation and inflexibility have led in recent times to the closure and amalgamation of structural subunits and the introduction of matrix and temporary structures for greater flexibility. These will be examined in more detail below.

Experimental facilities

Most institutes have some sort of workshops for the fabrication of their own experimental apparatus, particularly necessary given the difficulty involved in obtaining equipment from outside manufacturers. In this section, however, we shall devote attention to the increasing, although it is considered still inadequate, number of Soviet institutes which have what the Soviets call 'experimental' (*opytnye*) facilities. This refers not to laboratory facilities for doing scientific experiments, but rather to facilities for the construction and testing of prototypes. As has already been mentioned, relatively little Soviet R&D is done within enterprises. Instead it is done in independent institutes of the Academy and branch networks. This has produced a serious problem of a gap between research and production, a gap which leaves no place for the testing and adjustment for production conditions of new products and processes. Enterprises are often only interested in using their facilities for regular series production, and have neither the capacity nor interest in working on the testing and development of new products. Institutes traditionally have not had testing facilities, with all their resources being devoted to research. The resulting problem of *vnedrenie*, the introduction of the results of research into practical application, is seen as one of the most serious facing Soviet industry. A GKNT department head wrote in 1979 that over half the new products or processes offered to GKNT by the Academy of Sciences or *vuzy* could not be used because of design or development shortcomings, mostly caused by lack of experimental facilities.[43] Analysts consider that the best *vnedrenie* is achieved from work done in institutes where the proportion of those involved in research to those involved in experimental work is at least 1:1.[44] This does not apply to many institutes. In the late 1970s the proportion was roughly the same.[45] In the most recent data over one-third of research institutes and design bureaus in the machine-tool sector have no experimental facilities.[46] The Academy of Sciences, particularly the central Academy, has always been particularly badly off, especially since the removal of its more applied institutes in the early 1960s.[47] In the late 1970s, while the Ukrainian Academy had 64 production facilities able to build and test prototypes for its 77 institutes, the central Academy had virtually none.[48]

Efforts have been made to build up the testing facilities of research institutes. Within the branch sector, however, ministries have been very reluctant to provide funding for experimental facilities within their institutes. Thus in the late 1970s it was considered that 2–3 per cent of total capital investment in the production sphere should be devoted to building up experimental facilities. But expenditure levels of only 0.4–0.6 per cent were set in the plans of that period, and many of these allocations were not spent.[49] Indeed one source claims that investment plans for experimental facilities were being fulfilled to only 60–70 per cent.[50] By 1985 the situation had not changed, with still only 0.5 per cent of total capital investment going on experimental facilities.[51]

Only in very recent times has the central Academy begun to build up its experimental, testing, and prototype production facilities. There is the usual talk of the Academy being given the poorly performing enterprises of industrial ministries, particularly those in central locations which face severe redevelopment restrictions.[52] However, ministries are usually extremely reluctant to hand over even the most unprofitable of facilities, while another attempt to get facilities 'on the cheap', by having the Academy take control of the production facilities included in the new MNTKs, has also been abandoned. Consequently it appears that the Academy is being forced to undertake a major building programme of its own.[53]

It is the republican Academies which have led the way in obtaining experimental facilities. They rely to a far greater extent than the central Academy on contract payments, and therefore good industrial R&D, for their funding. One of the pioneers was the Siberian department of the Academy, which developed the so-called '*vnedrenie* belt', examined in Chapter 3. Under this approach industrial ministries built, equipped, and funded developmental facilities in Novosibirsk, with the relevant Siberian Academy institutes providing the scientific input and management. Despite other republican Academies following suit, the '*vnedrenie* belt' has been generally recognised as a failure, because of the ministries' habit of taking over the facilities and using them for their own production plan fulfilment purposes.

Therefore a number of republican Academies, led by the Ukrainian, have started to set up their own experimental facilities. They are usually subordinate to a research institute. A complex consisting of a research institute, a design bureau, and experimental facilities is called an 'engineering centre' or an academy scientific–technical complex (or association) (ANTK or ANTO). Besides the Ukraine the most publicity has been given to those in Belorussia. There they have set up special design–technological bureaus (SKTBs) with experimental factories in the institutes of the Academy's Departments of Physics and Mathematics and Physical–Technical Sciences, plus a Central Design Bureau serving the whole Academy. In 1982–4 these facilities were responsible for 92 per cent of the contract work done in the Academy, and it is mainly down to them that in 1980 the average yearly value of each contract project was 56,000 rubles, two and a third times higher than the 1970 figure.[54]

Administrative departments

Most Soviet research institutes are large enough to require quite extensive administrative apparatuses. It will be remembered from above that 214 of 791 departments in 44 Kiev research institutes were concerned with non-scientific matters. They dealt with such things as technical servicing, foreign relations, personnel, planning, accounts, raising qualifications (extension training), organisation of exhibitions, the scientific organisation of labour, civil defence, legal matters, and patents and licensing.[55]

The functions of the various departments should be clear simply from their name, although it is worth mentioning that the planning and financial departments have particularly extensive tasks in a centrally planned system, tasks with which it is said they are increasingly unable to cope.[56] The Personnel Department is also considered to have an important political role in the vetting of potential staff for political, ideological, and often even racial suitability (i.e. the maintenance of Jewish quotas and the right Russian/non-Russian balance in Central Asia).

Important political control work is also done by the first or special department. Although formally subordinate to the institute director,[57] these departments are generally considered to be in the KGB hierarchy and staffed by KGB officers. The first department is responsible for the maintenance of security, in the usual sense of the issuing and checking of passes, registration and securing of classified documents, etc., although the extent of such procedures might perhaps be seen as unusual and excessive in Western eyes. But the first department also has more directly political functions. For example, Zemtsov writes of the questionnaires and references on political reliability that had to be submitted by the staff of the Institute of Applied Social Research (IKSI) to that institute's first department, while it also sent its inspectors to dissertation defences, meetings, and conferences.[58] It should be noted that IKSI was a particularly controversial and politically sensitive institute. Raisa Berg writes in her memoirs how, as a researcher in the

Institute of Cytology of the Siberian Academy of Sciences, she was called to the special department of the Siberian Academy's Institute of Automatics to sign a denunciation of the noted economist Agenbegian for his criticisms of the state of the Soviet economy. The Institute of Cytology did not have a special department of its own.[59] The Institute of Automatics is of course militarily more sensitive than the Institute of Cytology, suggesting that the special department's functions are primarily although not solely concerned with security in the usual sense.

'Social' organisations

The last units in the institute which need to be mentioned are the so-called 'social organisations', the trade union, komsomol, and party organisation.

Soviet trade unions enjoy virtually 100 per cent membership among the working population. The unions are organised on an industrial basis, with all institute and university staff being members of the same Union of School, Higher Education and Scientific Workers. It surely hardly needs to be said that Soviet unions do not see it as their function to oppose management in an aggressive or conflictual way. Indeed many outside observers would claim with some justification that they work together with management to impose strict working regimes on workers. It is certainly true that trade unions work closely with management and in an almost entirely subordinate position. This role is best exemplified in the case of science by the presence of trade union representatives on 'efficiency review' commissions of each research unit. The efficiency review commissions are clearly and openly used to enforce discipline and remove inadequately performing workers from institutes. It is also not unknown for them to take action against researchers who have simply earned the displeasure of management for reasons that have nothing to do with efficiency (see Chapter 8). There are no indications that union representatives take an anti-management or pro-staff position in the commissions' work.

For the staff of institutes and other Soviet working collectives the main focus of the unions' work is their social welfare activities. The unions are responsible for overseeing working conditions and providing a wide range of social amenities. Many amenities such as childcare, medical care, rest and recreation, housing, and even food (through the 'subsidiary farm' and 'food order' systems)[60] are provided by the work collective. Clearly the extent to which such amenities are provided by work collectives depends on the size and prestige of each collective. In 1987 the prestigious and highly profitable Institute of Eye Micro-surgery, under S.N. Fedorov, spent 1.3 million rubles on 'social development', primarily on a leisure centre outside Moscow and a children's camp on the Black Sea, out of a profit of 3.2 million.[61] In the science sector many such facilities, particularly health care and rest and recreation facilities, are provided by umbrella organisations, such as the ministries or the Academies.

As far as working conditions are concerned it is not unknown for unions to take an interest in improving conditions. Lower-level representatives are not professional union organisers, but rather rank-and-file workers 'taking their turn'. They could well be prepared to take up relatively minor problems, such as heating, rest-rooms, etc., with management. As far as social amenities are concerned, the extent to which they are available depends very much on the energy and 'pushiness' of the responsible trade union officials, while the justness with which they are distributed depends on those officials' fairness and honesty. Those qualities vary enormously in each individual case.

The komsomol is the youth organisation of the Communist Party, recruiting members between the ages of 14 and 28. Membership among university students is virtually universal and effectively obligatory. It drops off in institutes. Komsomol organisations have primarily propaganda and political and social control functions.

I have written extensively about the party organisations of research institutes and universities elsewhere,[62] and shall do no more than provide a very brief summary here. Each working collective in the USSR with three or more party members has a primary party organisation, made up of all the party members of the collective. The data are very unsatisfactory, but I have estimated elsewhere that something like 20 per cent of all staff of research institutes and perhaps 30 per cent of research workers are party members.[63] On average something like 60 per cent of heads of departments, sectors and laboratories, in both the Academy and branch sectors, are party members.[64] It is safe to assume that virtually all institutes have their own primary party organisation (PPO). Depending on its size the PPO has as its executive a party secretary working on his own, or more usually heading a party bureau or party committee in the largest institutes. The party bureau or committee includes in its membership the director of the institute and other senior managers, the trade union chairman, and the komsomol secretary. The task of the PPO is to provide political leadership in the institute, through the exemplary work of party members, and to assist management in analysing and improving work practices. It can take party disciplinary measures against poorly performing party members, recommend administrative measures against non-party members, and has important powers, summed up in the 'right of control', to check and correct the work of management. The PPO would seem to have the potential to be, to quote one Western commentator, the Communist Party's 'ultimate device of control and management'.[65]

However, the evidence suggests that in fact the party organisation, and more particularly its secretary, operate 'in the pocket' of the institute director, and in most cases are unable or unwilling to oppose him.[66] Clearly the PPO is a body that cannot be ignored by Soviet researchers, least of all by the substantial and important minority who are party members. But in general it acts as a reinforcement of the control exercised over the workforce by management.

Summary

For a rank-and-file researcher undoubtedly the most important person in his working life is his immediate superior, the laboratory head. He will determine, quite probably in a detailed way, what work the researcher will be doing on a day-to-day basis. For the laboratory head, and ultimately for all the institute's staff, the most important person is the institute director, perhaps via his deputy for science. These senior scientists do the actual managing of the research process, and are only assisted by the various bureaucrats, the most important of whom is the scientific secretary. All researchers will be well aware of the existence of the various control agencies – the party organisation, the first department, the trade union. These agencies will make their presence felt particularly at times of political tension – a purge of dissidents, a campaign against Sakharov, a crackdown on foreign contacts. Essentially they are seen as reinforcements of the regular management structure, not a source of 'alternative' management. To that extent it could be said that the Soviet management structure is essentially the same as the traditional Western one – researchers working in laboratories and reporting up through a vertical hierarchy to an institute director.

Non-traditional approaches

Of course the 'traditional' Western model of regular and fixed laboratories reporting up through a single vertical line of command is hardly universally adopted in Western science, something the Soviets have become increasingly aware of in recent years as they try to break down the rigidities of their traditional system. The primary approach has been to unite science and production organisationally in NPOs and MNTKs and thereby reduce *vnedrenie* problems. These will be described in a moment. But there has also been increasing interest in project rather than institution-based planning and management, matrix structures and other forms of dual subordination, temporary laboratories and research groups, and scientific cooperatives.

Project and matrix structures

I have written elsewhere about project planning and management in the USSR,[67] and provide some details, particularly on planning aspects, in the next chapter. The issue is becoming topical again as new policies on *khozraschet* and self-financing are accompanied by statements that funding will no longer be provided on a 'block' basis to institutes but by research project. However, it would appear that, although there has been an accompanying campaign to rationalise institute structures in the sense of cutting down on the number of subunits, the project funding approach will not produce any significant change to traditional structures.

At times in the past, particularly in the mid- to late-1970s, there was talk of matrix structures being introduced as part of a major project-oriented research planning and management system. Such structures involve members of research institutes retaining administrative subordination to traditional department heads but at the same time joining research teams responsible for particular products or processes. The inspiration for such an approach seemed to be directly from US experience. It has apparently been formally implemented in the NPOs of the chemical industry and in some defence industries. A widely publicised experiment was also introduced into the R&D units of the huge Uralelektrotiazhmash (Urals heavy electrical machine building) plant in Sverdlovsk.[68]

However, the approach has received considerably less publicity in recent times, and one suspects that the difficulties of allocating rights and responsibilities within the matrix organisations and the determinedly institutionally based planning and resource allocation system have precluded its broad formal application. It remains to be seen whether any shift towards project funding will lead to a reactivation of interest in such an approach.

Temporary structures

A newer approach which is at the moment attracting considerable attention is temporary research structures. They come in two forms – temporary laboratories within existing permanent institutes and 'temporary creative collectives', which exist as essentially independent organisations. They both have a number of common features: they are permitted to exist for only a fixed period of time, between three and five years; they are established for the solution of a particular task, usually on a contract basis to an enterprise; and they are designed to provide the opportunity to earn reasonably substantial amounts of money through generous bonus schemes and temporary appointments to higher posts.

Temporary laboratories within existing institutes have been set up in both the Academy and branch sectors. The first such laboratories were established in the Academy in the late 1970s in the Institute of Applied Physics in Gorky, the Leningrad Physics-Technical Institute, and some institutes of the Urals Science Centre.[69] These laboratories were seen as successful, with the lead times of their R&D work cut substantially.[70] They were given formal recognition in a GKNT–Academy decision to set up eight temporary laboratories in the Academy, five of which were in the Department of General Physics and Astronomy.[71] It seems probable that these eight included the pioneering laboratories just mentioned. By 1987 permission had been given for 40 temporary laboratories in the Academy, of which 31 had already been established and nine were on the way.[72] The new Temporary Statute for Academy institutes approved in 1987 gave the laboratories full legal status.[73] The Academy laboratories are still seen as a success, with, for example, one in the Academy's Scientific–Technical Association (the Academy's main instrument-making facility) supposedly cutting the development time of an apparatus for drawing fine-film structures from the expected six years to 14 months.[74] All the indications are that it is the bonuses promised in advance that bring about such extraordinary improvements in productivity. Indeed academician Alferov, the director of the Leningrad Physics-Technical Institute, admits that one of the problems with the laboratories is that those staff members who have not gained appointments to them resent the money being earned by those that have.[75] The other problem mentioned in connection with the laboratories is that given that they usually work on *khozraschet*, i.e. a profit-making basis, the customers for whom the work is done tend to expect excessively short-term results.[76] One expects that if it is true that the laboratories go out of existence at the end of their statutory lifespan, even if their goal has not been achieved, and that in such a case the staff lose their bonuses, the pressure for short-term results comes not only from the customers.

The temporary 'brigades' in the branch R&D sector appear to work in much the same way as in the Academy. In the early 1980s six branch institutes in Leningrad each had between five and 25 such brigades, usually for non-planned urgent tasks.[77] This suggests funding from special funds, and probably again good money to be made by those fortunate or competent enough to gain appointment to the laboratories. A 1985 article describes how the Leningrad association 'Lenneftekhim' (Leningrad Petrochemicals) was directed by its ministry to do the R&D work needed to build a gas condensate refinery producing petroleum, diesoleum, and avgas, a surprising directive given that the association had never worked with gas before. A temporary creative brigade of researchers and designers was set up, and the research was completed on schedule in three months and the design in 12 months.[78]

'Temporary collectives' differ from temporary laboratories in that, although they are usually based in an existing institute, they are, theoretically at least, able to hire personnel from more than one permanent institution. Like the laboratories they are set up for a fixed period to work on a specific task. They are funded out of GKNT's reserve fund, with bonuses again being a large component of remuneration. People employed by them have their jobs kept for them in their home institute. The first such collective was 'Start', set up in April 1985 to develop a new generation of computers based on parallel architecture and using the 'Kronos' microcomputer developed by students at Novosibirsk State University. It was headed by V. E. Kotov, while its deputy head for science was the son of Gury Marchuk, the president of the Academy of Sciences. It employed about 155 staff, of whom 50 were from the Computing Centre of the Siberian Academy and the rest were from the All-Union Academy's Computing Centre in Moscow, the Institute of Cybernetics of the Estonian Academy, and the Severodonetsk NPO 'Impul's' (Impulse) of the Ministry of Instrument Making. Given the difficulties of housing and residence permits, 'Start' had no single geographical centre, although its primary base was in the Siberian Computing Centre. It reported management problems, since it was not allowed to have any administrative staff (although some administrators did masquerade under scientific appointments). Complaints were also made of outside bureaucratic interference, particularly in its hiring policies.[79] 'Start' was closed down in 1988, since its task had been completed, with it being stated that a scientific–technical complex of Academy and branch institutes and major factories was required to move to the production stage.[80] Meanwhile, the temporary collectives have been given a legal statute, and by 1988 ten were in existence.[81]

The temporary structures of both kinds seem to answer a shortcoming of the traditional Soviet R&D structure, and indeed seem particularly suited to one of the supposed advantages of the Soviet system – the ability to prioritise and readily move resources around. But it is an approach which might appear to be suitable only for a relatively small number of high-priority tasks, with the best and most highly motivated researchers able to work hard together and earn substantial rewards. There are those who would like to see the approach applied to all R&D, while even current enthusiasm for the broad application of the Western research grant system shows a general commitment to such an approach.[82] All this would imply a very much leaner R&D sector than is in existence at present. One doubts that as yet there is the commitment to such radical change. In the meantime of course the temporary approach produces resentment among those left outside the system, and as a consequence possibly leads to a decline in morale, commitment, and productivity in the 'traditional' bread-and-butter sectors.

Scientific cooperatives

Cooperative in Soviet parlance refers to a number of people joining together to work in their spare time outside the centralised planning and management system. A number of R&D cooperatives sprang up in the Soviet Union in the late 1960s, the most famous of which was 'Fakel' (Torch) in Novosibirsk. The cooperatives were usually made up of a number of young researchers with cross-disciplinary skills who would work in their spare time, usually using the facilities of the organisations in which they held staff positions, on short-term contracts for a range of customers in the R&D or production sectors. The tasks were usually writing new computer programs or debugging old ones, developing or improving instruments, and doing various forms of consulting work. The cooperatives met with considerable market-place success but less bureaucratic success. Despite the support of local scientists and komsomol and party organisations, 'Fakel' was eventually closed down by the Ministry of Finance, on the grounds that the members were contributing to inflationary pressures in the economy by making too much money! Some commentators have suggested that the emnity of branch institutes which had been shown up as incompetent by the cooperatives also contributed to their demise.[83]

With Gorbachev's promotion of cooperatives throughout the economy, R&D cooperatives have begun to appear once again. They have again faced opposition from the Ministry of Finance and some 'establishment' scientists, but in the present circumstances are able to get their own way.[84] Some have grown into quite substantial organisations, with for example the 'Engineer' cooperative based in the Riga Institute of Civil Aviation Engineers setting up a network in a whole series of institutes, design bureaus, and enterprises.[85] The Moscow Centre of Youth Scientific and Technical Creativity (NTTM), which is paralleled by similar bodies in a number of Soviet cities, had attracted orders worth two million rubles within six weeks of being set up. Some examples of the requests made of it are: 'to develop an automated system of operational control over pipeline construction projects'; 'to develop a transfer line for gilding bracelets at a jewellery plant'; 'to perform structural analysis of the magnetohydrodynamic generator magnet for the Riazan thermal power station'; and so on. The Centre has a Scientific–Technical Council which examines these requests, seeks out groups of specialists from throughout Moscow able to take on those that are deemed feasible, and then draws up contracts between the customer and the group. The Centre is apparently not connected to any particular institution, and therefore faces difficulties with access to facilities and materials.[86]

Those difficulties are clearly going to be less for cooperatives based in a single institute or bureau, although even here one imagines that the bureaucratic difficulties involved in regulating access are substantial. At the moment the top-level commitment to the cooperatives is strong enough that, for some, these problems could be overcome. If that is so the cooperatives might bring benefit to those areas of research, development, and innovation (RDI) that have traditionally been most neglected, including debugging new products and processes. One doubts, however, that even the most enthusiastic young specialists working in their spare time, operating in the unfamiliar and probably hostile environment of production enterprises, will do much to solve the USSR's innovation problems. However, the more specialised institute-based cooperatives will again suffer less from those difficulties.

Joint research–production facilities

As has just been suggested, one of the most serious problems that most commentators have seen in the Soviet R&D process has been that of *vnedrenie*, i.e. the introduction and assimilation of research developments into production. It has not been the purpose of this book to study such issues in great detail. Many other Western commentators have done so, often working from the assumption that the problem is in the production sector, or in the link between the production and research sectors, rather than in the research sector. Particularly striking has been the separation, in administrative, planning, incentive, and even geographical terms of research and production facilities. This separation has a long tradition in the Soviet Union. Nevertheless in recent years the authorities have worked hard at overcoming it, through the establishment of various joint science–production institutions. They are of importance to this study because they have had or potentially will have major effects on the way research itself, not just innovation, is carried out.

The first major effort was the science–production association (*nauchno–proizvodstvennoe ob"edinenie*, NPO).[87] The basic idea of the NPO is that a research institute and a production organisation are combined together in a single administrative structure. Ideally design and developmental facilities should also be included, producing an organisation able to cover the whole R&D cycle. There have still been unresolved differences of opinion among planners, managers, and commentators as to what extent NPOs should be responsible for series production of the products they develop. NPOs are related to production associations (*proizvodstvennye ob"edineniia*, PO), in that POs also include R&D and production facilities, but in this case a production enterprise heads the association, whereas NPOs are headed by a research institute and its director.

The NPOs made their first formal appearance in the September 1968 decree on the reorganisation of science and technology. As with so many of the other measures announced in that decree, NPOs were first set up on an experimental basis in the Ministry of Electrotechnical Industry. The experiment worked well and NPOs rapidly increased in number throughout the industrial sector. By early 1975 there were 97 NPOs in operation. At the end of that year a formal NPO Statute was published.[88] Since then there have been ups and downs in the spread of the NPOs, but the general trend has been very much up. Under Gorbachev there has been strong support, with calls from top leaders for a rapid increase in their number in the next couple of years, brought about by amalgamating all branch institutes except those classified as 'head' institutes into either NPOs or POs.[89] By early 1988 there were nearly 500 NPOs in operation. This was twice the number in existence in late 1985.[90]

The NPOs have indeed been one of the few organisational innovations of the Brezhnev era that have maintained their momentum and have been generally considered as a success. Data are presented to show that they radically cut lead times in the introduction of new technology and products, and that the technology and products developed by them are generally of higher quality and yield higher returns.[91]

But this is not to say that the NPOs do not face problems. Many appear to have been put together 'formalistically', with inappropriate institutions being combined or organisations scattered over an enormous geographical area. There are strong hints that at times ministries have deliberately put their weakest factories into NPOs in order to reduce their responsibility for them.[92] These longstanding problems are obviously still troubling the NPOs. Shkabadnia, until recently the Minister of Instrument Making, heading a ministry which is always keen to be at the forefront of organisational reform, proudly announced in early 1986 that they had already amalgamated all their R&D organisations into NPOs and POs. However, he admitted that sometimes it was difficult to get the right territorial and functional balance in their makeup, and that it would probably be necessary to look again at some of the new associations. A May 1986 article described the destructive 'musical chairs' that resulted from the ministry's haste.[93]

NPOs have also faced problems of poor management, since the management of the institutes which head the NPOs is simply not up to the complexities of managing and coordinating different units working in different fields. Shkabadnia in the article just referred to comments that they had been forced to establish POs in a number of cases where an NPO seemed to make more sense because of the lack of management capacity in the relevant institutes.

NPOs have also suffered from the usual Soviet problem of lack of developmental facilities. As we have already seen there is a serious resource allocation imbalance in favour of research and production at the expense of development in the USSR. In 1987 25 per cent of NPOs had no experimental-testing facilities at all.[94]

The situation is not helped by the overwhelming tendency of ministries to overload what developmental, testing, and prototype production facilities NPOs do have with planned production tasks, more often than not producing products that have nothing to do with the products being developed by the NPO. In 1984 the experimental factory of the 'Burevestnik' NPO in Leningrad had 90 per cent of its capacity loaded with extraneous production tasks.[95]

This situation is made possible by the fact that to all intents and purposes the constituent units of NPOs are administratively integrated in name only. In most NPOs the constituent units maintain their legal independence. In these circumstances the constituent units are subject to differing administrative subordinations, planning procedures and indicators, accounting procedures, and wage and bonus schedules. This has created enormous difficulties of coordination for NPO management, made it easier for ministries to deflect resources and capacity away from the stated goals of the NPOs, and led to recruitment and personnel management problems.

Under Gorbachev there has been concern to address some of these problems as part of the campaign to increase the number of NPOs. There has been much talk of the preparation of a new Statute.[96] It appears that the main aim is to establish truly integrated planning and management procedures within NPOs. Some moves have already been made in this direction, although they are moves that continue to be frustrated. A new methodology on NPO planning procedures, despite apparently having been drawn up in draft form by Gosplan, has still not been published.[97] On an experimental basis NPOs in five branches have already been moved over to integrated planning under the 'Science and scientific services' plan head. This means that the NPOs' production facilities cannot be subjected to production plan tasks. However, according to the commentators reporting the experiment it could also mean that NPOs will lose their production facilities altogether as ministries transfer them out of the NPOs rather than lose their production capacity.[98] In another experiment in the Ministry of Agricultural Machine Building NPOs were divided into two categories. Those which had a relatively small production capacity were placed entirely under the 'Science' planning head. This represented 15 of the ministry's 25 NPOs. The rest were placed entirely under the 'Production' planning head. In planning terms this would seem to turn them effectively into POs. However, the ministry then discovered that the transfer of even the limited production capacity of the 15 NPOs placed in the 'Science' category meant the loss of 80 million rubles worth of production. It was therefore decided to transfer *all* NPOs entirely into the 'Production' category.[99] Clearly the ministries are not going to give up their production capacity without a fight.

Another reorganisation of the NPOs which has attracted some attention in the last year or so is the setting up of All-Union NPOs (*Vsesoiuznye NPO*, VNPO). These are very large NPOs covering a whole product area and with major series production facilities. They are seen as best suited to high-technology fields in which scientists can usefully maintain control over their products right through the production stage. The best known is the VNPO 'Kriotekhnika' (Cryogenic technology), based on the highly successful NPO 'Kriogenmash' (Cryogenic generator construction) and including a number of other NPOs and factories working in the low-temperature engineering field. The founder–director of the VNPO, academician V. P. Beliakov, gained considerable publicity for his organisation and his calls for more such VNPOs to be established.[100] However, Beliakov died suddenly in August 1986. His successor, V. E. Kurtashin, has continued to push for the development and legal regularisation of the VNPO concept,[101] but he appears not to have the clout of his predecessor and VNPOs are now receiving less attention.

The history of the NPOs shows how difficult it is to overcome bureaucratic opposition, particularly from the branch ministries, even to a successful administrative reorganisation which has the full support of the top political authorities. It appears that the NPOs will continue to be the focus of branch R&D effort despite this opposition, but there is also increasing attention to forms of R&D organisation that are designed to get around the 'departmental' jealousies and production priorities of the ministries.

Inter-branch scientific–technical complexes

One such approach has been the inter-branch scientific–technical complexes (MNTKs). This organisational form is designed not only to overcome the problem of the separation of science and production, but also to attack that other traditional problem of Soviet economic management, *vedomstvennost'*, i.e. the 'sectional' rivalries of different institutional actors. The MNTKs could very roughly be described as inter-branch NPOs, since they include organisations from a number of administrative hierarchies. They are intended to have a serious production component, having real factories within them. Their main purpose, however, is to overcome the 'sectional' barriers erected by the branch ministries against the transfer of technology.

The Ukrainian Academy can take the credit for the development of the concept. The Paton Institute has been described as the model for the MNTKs (by Kudinov, a deputy chairman of GKNT and formerly party secretary of the Paton Institute),[102] which is perhaps a little unfair to the administrative agency of the Ukrainian Academy in L'vov, the Western Science Centre, which for over a decade has been running together with the L'vov regional party committee MNTKs based on voluntary cooperation and contracts.[103] The Ukrainian MNTKs were praised by Gorbachev at the June 1985 conference on science and technology, with the Politburo then discussing a decree on the universalisation of the complexes in December.[104] The decree has not been published in full, with only a brief summary appearing in *Izvestiia*.[105] I have also not had access to the Basic Statute of the MNTKs, apparently issued as a decree of the Council of Ministers on 31 July 1986. The original Politburo decision referred to 16 MNTKs, but there are now at least 23, working in fields ranging from anti-corrosion treatment to optical fibres. The original MNTKs were given one million rubles each by GKNT in early 1986 to start them off, and Gosplan approved an impressive list of construction projects for them in December 1986.[106]

The complexes are headed by a major research organisation, supported by a number of other institutes, design bureaus, developmental facilities, and production enterprises from a range of different branches and ministries. The production enterprises can be very large, with for example Uralmashzavod being involved in 'Ore recovery'. The constituent units do not devote all their energies to MNTK tasks, however. Thus the 60 per cent of their capacity that some Minpribor factories devote to the 'Machine reliability' MNTK is seen as a high proportion.[107] It is also said that the head organisation can bring in other non-constituent organisations if a particular need arises. Judging from a recent article on rotor technology, it appears that factories can be put under the 'sponsorship' (*shefstvo*) of an MNTK during the period that new technology for which the MNTK is responsible is being introduced into the factory.[108] It is stressed that the make-up of the complexes is determined directly by the Council of Ministers and that membership for those chosen is obligatory. The constituent units are all represented on the MNTK Council, which on the basis of the suggestions of the head organisation draws up integrated five-year and annual plans for all components of the complex, from basic research through to production, and covering funding, labour, and supplies.[109] The plans are then confirmed by GKNT, in agreement with the Academy, which is responsible for determining the long-term value of the planned work, and Gosplan, responsible for making arrangements for production. The plans of the MNTKs are included in the Science and Technology Section of the Five-Year Plan under a separate MNTK category. It is claimed that this will give them priority in funding and supplies.[110] In the day-to-day operations of a complex the decisions of the head organisation are obligatory for constituent organisations.[111] An attempt is being made to get away from the problem that plagues NPOs – that their experimental facilities are overloaded with production tasks – by putting the MNTKs' experimental (*opytnye*) facilities under the 'Science and scientific ancillary services' section of the state plan. That is, they will not be given basic production plans.[112]

Some MNTKs are supposed to cover the whole research–production cycle, but most seemingly are only expected to develop a product to the stage that it can be taken over by series production factories in the normal way.

A number of comments can be made about the MNTKs. Firstly and most importantly, the constituent units of the complexes are not fully and permanently subordinate to the complexes, despite calls for them to be made so.[113] They retain their usual subordination to a ministry or other agency, and only a portion of their work is covered by the complexes' plans. A GKNT official has claimed in one interview that the general directors of MNTKs will have at their disposal some proportion of the capacity of series production factories.[114] However, one senses some scepticism on the part of the journalist interviewing him, and in a recent interview which could only be described as authoritative Marchuk (at that time still chairman of GKNT) seemed to concede that MNTK plans must be approved by the relevant ministries before being passed on to Gosplan and GKNT. Originally this requirement was seen as undesirable.[115]

The continuing primary subordination of the constituent units to various ministries and agencies has had precisely the effect that one would expect – the classic symptoms of *vedomstvennost'*. The head institutes have enormous difficulty getting the cooperation of those parts of the MNTK based in other organisations. The general director of 'Catalysis' described it best when he declared that 'our MNTK partners sometimes remind me of people who have been dragged out to a *subbotnik* [a Saturday 'voluntarily' given over to unpaid labour]'.[116] The general director of 'Ore recovery' also describes in worried terms the increasingly centrifugal tendencies evident in his MNTK.[117] He states that the MNTK can be managed only on the basis of one-off decrees and instructions from superior party and state bodies because of the reluctance of the ministries and agencies involved to cooperate, a statement that has been echoed by others.[118] This *vedomstvennost'* means that the MNTKs, which in the final analysis rely on the goodwill of the bureaucratic agencies operating the constituent units, do not get the priority in funding, supplies, and facilities that they are supposed to get. The original proposals that the head institute of the MNTK be given direct and unqualified control of all resources earmarked for the MNTK were apparently vetoed by the Ministry of Finance.[119]

Other commentators point out that it is not only the subordinate constituent units that are liable to give or be forced to give first loyalty to their regular bureaucratic hierarchies. The head organisations of the MNTKs also have regular administrative subordination to ministries or the Academy. This is likely to mean that they also retain their traditional reluctance to engage in inter-branch cooperation. One of the head organisations is ENIMS, the institute most notorious for jealously protecting its own technological hobby horses and those of its ministry. There is nothing in the structure of the MNTKs to suggest that head institutes will be forced or encouraged to change their behaviour. A staff member of one of the branch NPOs involved in the MNTK 'Biological engineering' complained that the general director of the complex, the late Iurii Ovchinnikov, spent all the complex's money on his own Institute of Bioorganic Chemistry, the MNTK's head institute.[120]

The next problem to be discussed is no more unexpected than that of *vedomstvennost'* – it is the old one of getting the new products and processes into production. If it is true that the MNTKs are responsible for a product only up to series production, then they will face the problem that no matter how perfectly developed their new product is, enterprises and their ministries are unwilling to take on its full-scale, or even small-scale, production. The late Beliakov, of the NPO 'Kriogenmash', saw this as a problem at a very early stage.[121]

The other problems described in the press are also the sort of problems one would expect. Firstly there is the danger that the plans of the MNTKs are no more than compilations of work that the constituent units were already intending to do anyway. It is said that the Ministry of Non-Ferrous Metallurgy has simply reproduced nearly all the ministry's R&D tasks in the metallurgical engineering sector for the MNTK 'Metallurgical machine building'.[122] In such a situation the MNTK loses its strategic focus and is likely to fall into the old trap of *melkotem'e*.[123]

Perhaps one of the most interesting problems the MNTKs face is in the patent scepticism and lack of enthusiasm of those most closely involved. A GKNT official responsible for MNTKs expressed his frustration early in 1987 over the lack of interest of MNTK executives in establishing their legal status and spending available funds. He wrote that GKNT had expected an avalanche of documents from MNTKs once their Model Statute (*tipovoe polozhenie*) was approved in mid-1986. But only about half of the complexes had made any effort to take advantage of this opportunity to regularise their status by the end of the year. He also commented on their general reluctance to use the powers available to them. Only one MNTK was in any hurry to put forward proposals to spend the one million rubles allocated by GKNT, with 'Anti-corrosion' even having its money confiscated by Gosbank because of its failure to make use of it.[124]

The Academy of Sciences has shown itself to be if anything even less enthusiastic than the branch agencies. Right from the beginning there were complaints that academicians and the Academy's departments, with the exception of the Department of Machine Building, were unenthusiastic in their attitudes to the MNTKs.[125] Nevertheless the Academy was given a leading role in about half the MNTKs. However, they continue to display their reluctance to proceed with them. There were long delays in getting the MNTK 'Heat synthesis' into operation because of opposition within the Academy to the 'self-propagating high-temperature synthesis' (SVS) technology that the MNTK is responsible for. As a result the Academy's Institute of Chemical Physics was most reluctant to take over leadership of the MNTK. Its sister institute, the Institute of Physical Chemistry, is also a reluctant member of the MNTK 'Oil recovery', on the grounds the its field of work is not covered within the institute.[126] The Academy MNTKs have been singled out for delays in drawing up their draft five-year plans.[127] Many academicians probably see the MNTKs as deflecting attention and resources from the Academy's fundamental research.[128]

Whatever the attitudes of the Academy, its record with its MNTKs has not been a good one. Four of the Academy MNTKs have been transferred from the Academy to ministries. Another three have been placed under joint Academy–ministry supervision.[129] The most details have been given on the collapse of Academy management of 'Optical fibres'.[130] The Academy admits continuing problems with 'Industrial lasers'.[131] That would suggest that the only Academy MNTKs that have met with success are 'Scientific instrument making' and the two Ukrainian complexes, the Paton Institute and 'Powder metallurgy'.

Given that the Academy has long had the reputation of outperforming the ministries in any field, its failure with the MNTKs must be a serious blow for the new approach. Indeed the cynical scepticism of the bureaucrats that the MNTKs are a fad which will fade away might prove justified. Senior science managers such as Paton and Marchuk state their continued commitment to the complexes, despite admitted serious problems.[132] Others sound more sceptical, with one commentator in particular suggesting that the MNTK approach will be even more inappropriate and difficult to manage in conditions of full *khozraschet* and self-financing of enterprises.[133] The author states that the only successful MNTKs are those operating in fields that had already been working well. The best example is, of course, Fedorov's 'Microeye surgery', whose incredible empire is expanding into agricultural production and luxury floating hotel-hospital ships serving Saudi oil sheikhs, etc., on top of its current 400 operations per day on its Soviet production lines.[134]

The central authorities are committed at the moment to continuing with the MNTKs. It was reported in May 1988 that GKNT, Gosplan, Gossnab, the Ministry of Finance, and the Academy were preparing a number of proposals to strengthen their position. The proposals were to give the MNTKs status as legal entities, with their own accounts and bank balances. They would be able to form their own development, bonus, and foreign currency funds, financed on the basis of receipts from realised R&D work.[135] These proposals represent yet another and far from convincing effort to overcome the sectional jealousies and rivalries that have bedevilled the MNTKs.

When the MNTKs were first introduced it was possible to believe that, despite the problems that they were quite clearly going to face, their restriction to a small number of fields would allow the much-vaunted Soviet prioritisation to take effect. One now suspects that the central authorities have neither the will nor the power to enforce even these priorities.

Conclusion

In terms of formal structures within institutes the makeup of Soviet institutes does not differ greatly from that of traditional Western institutes. The top managers tend to be of scientific background, while the research units of the institutes are organised in a hierarchy of departments and laboratories. There are also the usual central 'functional' or 'staff' departments. At this level the primary differences are in the nature of planning and funding, in particular the almost total absence, in both formal and informal terms, of a project-oriented approach. This fact has considerable management implications which are considered in other chapters.

Soviet institutes also face the same sorts of problems as Western institutes: of optimal location, size, and vertical and horizontal spread of management charts, as well as the same sorts of problems of inter- and intra-disciplinary relationships in administrative structures. The Soviets have demonstrated a willingness to follow Western attempts to deal with these problems, particularly in the direction of project and matrix approaches. However, the adoption of such approaches has as yet been far from wholehearted.

Probably the biggest differences between Soviet institutes and those in the West, particularly in the applied field, are their separation from the production enterprises they are meant to be serving and the very considerable inter-sectoral jealousies that bedevil Soviet innovation. These problems obviously have an organisational expression, and attempts are being made to overcome them through organisational change – the amalgamation of R&D and production organisations, including across sectoral boundaries. However, such attempts have been less than successful because the problems are not purely organisational. They derive to a considerable extent from human nature, but also from the peculiarities of the Soviet planning system. One could conclude by saying that, although Soviet R&D structures are not perfect, the real problems of Soviet science and technology are not structural. The non-structural problems will become evident in the final chapters, in which we examine the conduct of Soviet research.

Chapter seven

The planning of Soviet science

Planning is regarded as one of the fundamental features of the Marxist–Leninist approach to socio-economic organisation, with the official attitude being that scientific research is susceptible to planning. By planning is meant not just the compilation and circulation of the research interests of researchers, but rather a 'directive' process in which superior authorities on the basis of a careful analysis of national needs and capabilities direct effort and resources in particular directions. It is recognised that the creative process has to be handled with care, particularly when fundamental research is involved. Nevertheless a high degree of planning is seen as an essential contribution to the efficiency and control of the scientific process.

Many scientists tend to be more cautious in their attitudes towards scientific planning. When planning was first introduced into the science sector in the mid-1920s there was considerable opposition from 'bourgeois' scientists who saw the process as an attempt at the imposition of political and ideological control over their work. Ever since, fundamental scientists in particular have expressed concern about the stifling of scientific creativity through excessive planning, while even applied scientists – indeed in some cases especially applied scientists[1] – complain about the negative consequences of interference by incompetent planning officials, the excessive bureaucracy involved in the planning process, and the lack of opportunity for rank-and-file researchers to contribute.

However, it would be dangerous to exaggerate scientists' opposition to planning of their work. As long as they have the opportunity to contribute to and even dominate the planning process – and as we shall soon see the 'scientific establishment' at least does have such an opportunity – opposition is muted and indeed planning is welcomed. It can provide a guarantee of funds and resources, and increase the pressure on producer organisations to make use of the work of the scientists.

History of the planning of science

It is only relatively recently that science planning has become a matter of priority, with extensive formal procedures being established. Thus science did not make an appearance in central economic plans until the late 1940s. In 1949 *Gostekhnika* (the rough equivalent of today's GKNT) prepared an annexe of special S&T projects for the annual state plan, something which apparently became a regular practice after 1953. However, these projects were not integrated with the plans of individual institutes, nor were they included in the five-year plan. Research tasks were not included in five-year plans until the Eighth Five-Year Plan (1966–70), when the first moves were made towards instituting the details of the present system.[2]

The Soviet planning system, even in its most extreme and idealistic forms, has never been designed to remove entirely the influence of lower-level operational bodies over the planning process. Indeed there tends to be something of a pendulum in the Soviet approach to planning, moving between centralisation and decentralisation. A highly centralised system produces serious problems of lack of incentive and responsibility at lower levels, as well as distortion and errors resulting from information overload. The pressure is then to decentralise planning responsibility, both formally and informally, to operational units. Informal decentralisation takes the form of relying entirely on the simple compilation of draft plans passed up from below or on previous performance to set targets ('planning from the achieved level'), of allowing higher degrees of cheating, and simply of giving those who have to implement the plans greater informal input. Formal decentralisation usually takes the form of reductions in the number of plan indicators and the introduction of profit-and-loss accounting (*khozraschet*). These developments, however, lead to the neglect of priorities, an overcommitment to the short term, and the increasingly glaring dilemma of what to do with organisations clearly unable to pay their way. The pressure then moves back in the direction of centralisation. Informal decentralising pressures are always evident in the Soviet planning system, as we shall see shortly. A major formal decentralisation campaign is also under way at present. It followed a period of centralisation dating from the late 1970s.

The planning process

The traditional planning system, which still forms the basis of Soviet S&T planning, consists primarily of five-year and yearly plans. Thus the Academy of Sciences has its five-year and yearly plans, as do all the branch ministries, and of course all the institutes in both sectors. The process begins with institutes preparing drafts of their own intended activities in the next plan period. They pass this draft to their administrative superiors – in the case of the Academy the relevant department and eventually the presidium (through the Planning Department of the presidium apparatus); in the case of the branch sector up the bureaucratic hierarchy of the ministry. These administrative bodies examine and collate the drafts, and add projects which they see as important or which have been passed down to them from higher authorities. The collation is presented, after checking by GKNT and Gosplan (and the Academy in the case of fundamental research done in the branch sector), as the draft Academy or branch plan, and the component institute plans are sent back down the hierarchies for possible negotiation. The plans are then issued as binding 'directive' plans.

The process is in fact a multiple one, since separate plans are required for the listing of projects, the allocation of funding, the allocation of supplies and materials, and capital investment. The finance and supply plans follow much the same procedures as the basic research plan, although different sections of the bureaucratic hierarchies are used (for example, the supply plans travel through the supply departments of the bureaucratic apparatuses and receive top-level checking by the State Supply Committee, Gossnab, rather than GKNT or Gosplan). Coordination between the plans is usually not good, with complaints being particularly common about the need to submit requests for supplies and materials before the research plan has been drawn up.

Despite the usual Western perception the most important plan is not the five-year plan. It tends to be quite highly aggregated and is essentially indicative. It is the yearly plan which contains detailed tasks and goals. A survey of *émigré* Soviet scientists and technologists found that 60 per cent considered the yearly plan the most important, 25 per cent the quarterly plan, and only 5 per cent the five-year plan.[3] The dominant importance of the yearly plan can be seen from the fact that efforts are usually made to round projects off to yearly periods. One source claims that over 80 per cent of projects are formally run over the calendar year. The most basic of the yearly plans is the research plan of the individual institute, known as the 'thematic' (*tematicheskii*) plan, and that is what we shall now concentrate on.

The institute yearly plan takes nine to ten months to put together and is made up of projects coming from a number of different sources.[4] These include national plans and programmes; the five-year and yearly plans of the particular branch the institute is situated in; one-off orders for research from higher bodies, in the form of decrees from 'directive' organs (these will often be inserted in the plan during the period of its fulfilment); orders from other institutes and enterprises from outside the ministry or agency in which the institute is situated; and 'exploratory' (*poiskovye*) or 'initiative' (*initsiativnye*) projects.

There are four main sources of funding for these different types of projects. Funding can come from the central budget, usually controlled by GKNT. The one-off 'directive' tasks, programme projects, and much basic research are likely to be funded in this way. Funding can also come directly from the budget of the institute's bureaucratic superior – the ministry or the Academy. This covers a large proportion of projects in the regular ministry and Academy plans. However, a considerable proportion of projects are also funded by direct contracts between the institute and user organisations. Finally, institutes have some discretionary funds, to be devoted primarily to exploratory or 'initiative' research.

We do not have good data on the breakdown between the different sources, particularly since it has changed significantly over recent years. These changes are particularly evident in the ratio of ministry versus contract funding in the branch sector, resulting from the introduction of the *edinyi fond razvitiia nauki i tekhniki* (single fund for scientific and technical development, EFRNT) in the 1970s (see below).

However, it appears that in general a goal is set in both the Academy and branch sectors to have between 20 and 25 per cent of projects set aside for 'exploratory' research, although particularly in the case of branch research the target is rarely met.[5] In Academy institutes contract projects can take another 25 per cent, higher in many union republic Academies, leaving about 50 per cent for projects allocated from central plans. In the branch sector the contract share is likely to be higher, although declining since the introduction of EFRNT. We have a breakdown from the early 1980s on the sources and funding of projects for the metallurgical, construction, and cellulose-paper industries, presented in Table 7.1.

There is room for the influence of the institute over all inputs into the planning process. As has already been suggested even projects allocated to the institute's plan by higher planning bodies are very likely to have originated as proposals made by the institute itself, with many projects simply being repeated year after year. As far as contracts are concerned institutes have a set percentage of their budget set aside for contract receipts, and so to some extent they are obliged to go out and seek business. In some cases the central ministry apparatus apparently determines who will order what research from what institute. One source claims that 70–80 per cent of contract work is the result of decrees coming down the ministry hierarchy.[6] But it appears often to be the case that institutes receive more requests for contract work than they need to fulfil that section of their plan. In that case they have freedom to choose.[7] 'Initiative' or 'exploratory' research is clearly the most free as far as the institute management is concerned, in that the funds set aside for such research can apparently be genuinely disposed of as the management likes. However, many commentators complain that not enough funding is set aside for such research, in either the Academy or branch sectors.

Table 7.1 Sources and funding of research projects in three industries, early 1980s, percentages

	Sources of projects			
	Central organs	*Ministry*	*Outside orders*	Initsiativnye
Metallurgy	24	53	7	16
Construction	20	20	50	10
Cellulose	11	65	22	2

	Funding of projects		
	State budget	*Contracts*	*Ministry*
Metallurgy	20	49	31
Construction	45	55	–
Cellulose	15	23	62

Source: V.I. Duzhenkov, *Problemy organizatsii nauki (Regional'nye aspekty)*, Nauka, Moscow, 1978, p.81

Hardly surprisingly, informal contacts and influence are important in the planning process. We have considerable detail on the efforts of the Georgian physicist Elevter Andronikashvili. He regularly turned to Igor Kurchatov, the highly influential head of the Soviet bomb programme, with requests for assistance in getting new projects off the ground.[8]

At other times Andronikashvili used party and government contacts, in particular his old friend Viktoriia Siradze in her various capacities as a district party secretary, a deputy chairperson of the Georgian Council of Ministers, and Georgian Central Committee secretary for ideology.[9]

One cannot underestimate the importance throughout this process of what we might call 'public relations' (although they are in fact sometimes quite private). It is important to sell oneself to the scientific community by producing articles, contributions to conferences, and students with good dissertations, and to the politicians by being politically loyal and willing to work on 'useful' projects, particularly in the defence sector. But it is also important to appeal to a wider audience. When Kurchatov was trying to give a spurt to fusion research in the second half of the 1950s, he followed the usual path of formally presenting a plan prepared by the Atomic Energy Department of the Academy of Sciences to the Central Committee (probably meaning here the Secretariat or even the Politburo). But he also adopted a public relations approach. His greatest coup was persuading Khrushchev to allow him to go to England to talk publicly about the Soviet fusion programme. This gave enormous impetus to what had previously been essentially a research programme, and turned it into a scientific–engineering undertaking. This required obtaining more resources from a wider variety of sources, and so in March 1958, during a Supreme Soviet session in Moscow, Kurchatov invited a number of industrial managers and Moscow and Leningrad party bosses to dinner to impress on them how important fusion was. Over the next two years, up to his death, Kurchatov 'established friendly contacts with many government leaders, took them on visits to the laboratories, and attempted in every way to popularise thermonuclear research'.[10]

Another popular form of public relations is the publication of articles in the political-popular press. I have been told by a participant in the Soviet laser fusion programme that he occasionally writes articles on the programme to be printed in the daily press over the signature of Basov, the head of the programme. Such articles sometimes put forward requests or make complaints. But more often than not they are neutral or enthusiastic accounts of progress so far. They seem to fulfil the same vague role as 'corporate image' advertising in the West.

Scientists also sometimes form alliances with writers, who popularise their views in fiction or publicist works. For example, Daniil Granin is said to have pushed Nikolai Semenov's views on fundamental research during the great Academy debate of the late 1950s.[11]

It appears that in the Soviet Union as elsewhere those who use public relations run the risk of becoming the victims of their own publicity. In the early days of Soviet fusion research an experiment was being carried out in Artsimovich's laboratory designed to demonstrate the existence of neutron radiation in powerful deuterium pulses. The existence of such radiation would be evidence that a fusion reaction had been obtained. Presumably it was expected that the experiment would be successful, since D.V. Efremov, the powerful Minister of the Electronics Industry, was there. But despite the best efforts of the researchers no effect was evident and the minister left disappointed. The researchers then presumably tried harder, because later they were able to demonstrate the required effect. This led to a flood of visits and congratulatory telegrams from government leaders and scientists. Despite the fact that Artsimovich himself had his doubts (the neutron radiation was accompanied by hard X-ray radiation, an unwanted effect), the experiment was taken as a demonstration of a radiant future for fusion, and in particular for the pinch approach. When Artsimovich's suspicions were later confirmed by US experimenters the future of fusion suddenly became cloudy.[12] One gets the sense here of the universal pressure researchers, desperate to maintain funding levels, can find themselves under to demonstrate good results to their paymasters.

Both the formal and informal aspects of the planning process provide for a degree of flexibility and input from scientists. However, from the discussion so far the process would seem to favour institute management and influential scientists. What is the situation as far as rank-and-file scientists are concerned? It varies so much according to the rank and motivation of the scientist and the type of institute in which he or she works that no generalised example can be devised. Planning is going to mean very different things to an unambitious rank-and-file researcher in a branch institute doing essentially a technician's job, an ambitious and capable researcher in the same institute, a theoretician in an Academy institute, or a member of a team working in a 'big science' field such as particle physics.

But the evidence suggests that influence over the planning process comes only when a person has obtained a management post, i.e. has become at least the head of a laboratory. At levels below that the rank-and-file researcher is simply given tasks by his superiors, with very little opportunity to provide input into the planning process.[13] Many Soviet researchers, treating science as 'just a job', are happy with such a role. Some fulfil it conscientiously and competently, but all too many do so with far less commitment (see Chapter 9).

At this stage the more creative and ambitious scientist, as long as he does not require particularly complex or expensive equipment, will usually find that the fulfilment of his planned allocated tasks still leaves time to do work of his own. If this 'private' work is successful and impresses his superiors he can hope to have it eventually included officially in the institute's plan.[14] This will allow him to be provided with formal allocations of funds and resources (which by no means guarantee him easy or adequate access to those resources). It could also well lead to him being put in charge of a laboratory centred around his particular project. (This means that Soviet institutions often are quite 'project-oriented' in their very structures, although it is an inflexible 'project' approach, since laboratories are more easily opened than closed.) He will now have personnel under him whose planned work will be based on his project. Once at this stage, the researcher should be well established at the 'feed trough', and can expect a relatively easy time well into the future.

In fact a number of commentators, both Western and Soviet, have suggested that Soviet researchers are better off under the Soviet system of science planning and funding than their Western counterparts. Western scientists rely above all on fixed-term research grants to keep their work going, forcing them to devote considerable amounts of time to preparing grant proposals and to ensuring, sometimes against their will, that their research is in 'fashionable' fields. The Soviet scientist, confident that his project will reappear year after year in the plan, does not have such concerns. While institute management and bureaucratic apparatuses in the Soviet Union devote considerable amounts of time to planning procedures, rank-and-file researchers spend their time on obtaining supplies instead. The reduced need to follow 'fashion' means that Soviet scientists are sometimes able to work away quietly and sometimes fruitfully in scientific 'backwaters'.

A number of Soviet commentators feel that the traditional system gives far too much say to scientists, and in particular gives too much weight to personal influence and established reputation and too little to the needs of the economy. For some this means that non-science management bodies have to become more involved.[15] Recently some senior Soviet science managers, dissatisfied with what they see as the slackness of their system, have called for the introduction of a tougher research grant system.[16] While the trend is vaguely in that direction – with the Soviets struggling to introduce project- rather than institute-based planning and funding – they are still a long way from reaching that goal.[17] It will presumably not do a great deal to reduce the weight given to personal influence and established reputation anyway.

One of the noticeable features of the existing system is that the main beneficiary is institute management. It is the institute director or laboratory head who has control over the allocation of funds and listing of projects in plans. The lack of individual research funding means that individual researchers have no choice but to rely on the goodwill of management. As we shall see in the next chapter this goodwill can often come at a high price and be withdrawn in a rather arbitrary way.

Plan implementation

There is further flexibility built into the scientific process at the plan implementation stage. The Soviet planning system in all sectors of the economy is characterised by a high degree of *korrektirovka*, i.e. the practice of changing plans during fulfilment to take into account changing priorities and needs, unforeseen supply difficulties, and the simple threat of non-fulfilment. One source claims that as much as two-thirds of R&D plan estimates are subject to correction, mainly in terms of transferring funds from one project to another.[18] This seems to mean that no change is made to the project itself, but simply to its funding allocation.

Data from the branch institutes of the Tatar Autonomous Republic show that adjustments are also made to the projects themselves, indeed to an average of 20 per cent of the yearly volume of work; adjustments here seemingly referring to the projects as listed in the 'thematic' plan.[19]

Recently data were provided on additions to the five-year plans of the Sectors of the Academy of Science during the Eleventh Five-Year Plan. The Physics Sector had its plan changed by 37 per cent; Chemistry by 15 per cent; Earth Sciences by 12 per cent; and Social Sciences by 61 per cent.[20] Very often such corrections are ordered by higher authorities or contracting partners rather than by the institute itself. Of 35 corrections to the 1981 plan of an institute in the Ministry of Heavy Machine Building, 23 were at the initiative of the contractor, 11 from higher authorities, and only one from the institute.[21] Corrections to items included in central plans and 'directive' documents naturally have to be approved by higher authorities. 'Exploratory' work funded by the institute can be modified at the will of the director. The legal situation seems also to leave room for institute managements to modify contracts unilaterally.[22] It should not be thought that all 'outside' corrections are necessarily impositions on institutes or unwelcome to them. Institutes are well able to encourage outside bodies to propose corrections on their behalf, particularly as it is often in the interests of these bodies that the corrections be made (especially the higher authorities who are responsible for ensuring plan fulfilment).

The final area of flexibility in the planning system is in plan reporting. Plans contain many indicators, some of which are clearly understood as being more important than others. Traditionally it has been well understood that the provision of a finished piece of successful research is one of the least important indicators in Soviet R&D plans. The most important indicator seems to be that the allocated funding has been spent in the plan period of the project, usually the straight calendar year. The effect of such a system is twofold: firstly, it means that plans are hardly ever formally not fulfilled; secondly, institute management is easily able to move funds around from project to project quite informally, and as long as the money is spent the plan will be recorded as fulfilled.[23] Attempts have been made over the last 20 years, and with particular urgency in recent times, to change the system so that plan fulfilment will be recorded, and payment made in the case of contract research, only for a completed piece of research adjudged likely to achieve a set amount of economic return once introduced into production. All such attempts so far have run into serious problems of how to evaluate the success of completed research, what to do about research which fails for genuine scientific reasons, and simply how to keep institutes that have never maintained high levels of operating funds going while the new systems are implemented.[24]

The new systems are particularly susceptible to the final area of flexibility in the planning system, i.e. cheating. This is because the new systems depend so much on measurement of economic returns from the results of R&D once implemented (usually expected rather than actual returns, since the timescales for payment would be too long in the latter case). The temptations for institute management and individual researchers to inflate return figures are obviously considerable and all too often not resisted. This sometimes entails what can only be described as the bribery of users to concur in the inflated estimates. One article refers to scientists working on a particular project who granted lucrative 'consultancy agreements' to the specialists who were to be involved in determining economic returns; another refers to a case in which a scientist included a production manager as co-author of a piece of R&D in exchange for an inflated return figure.[25] Sometimes credit is claimed for non-existent work. A 1986 newspaper article attacked an institute which contracted to develop a new construction technique for a construction trust and received a down payment of 15,000 rubles. According to the trust's chief designer they never saw anyone from the institute again, but did receive another bill when the contract period was up. The trust paid, because the funds had been allocated and therefore had to be spent. All the records showed an effective and profitable piece of R&D.[26] One of the simplest forms of cheating, designed to ensure trouble-free plan fulfilment, is the inclusion in the plan of projects that have already been completed.[27]

Planning reform

As mentioned at the beginning of this chapter, Soviet planning theory and practice alternate in their emphasis between centralisation and decentralisation. Planning reform tends to move in one or other of these directions.

Centralisation in the 1970s

We have seen that the traditional planning system contains a far greater degree of flexibility than one might have expected. That flexibility has been seen as counterproductive by many Soviet analysts and policy-makers, with some even prepared to go so far as to claim that the traditional planning system is no more than a formality, with institute management in fact being able to dispose of funds as they wish.[28]

In the late 1960s steps were taken to tighten up the system. The first approach was limited to the branch sector, and entailed the greater centralisation of research funding in a new ministry-level Single Fund for the Development of Science and Technology (*edinyi fond dlia razvitiia nauki i tekhniki*, EFRNT). Originally the fund was introduced into the Ministry of Electrotechnical Industry on an experimental basis. It only gradually gained acceptance, and was introduced into all non-defence industrial ministries in 1979. In the EFRNT system the bulk of branch R&D spending is derived from a single fund centralised in the central ministry apparatus. The money is distributed to R&D organisations, usually via a head organisation, according to a *zakaz-nariad* (work-order). This is a type of contract drawn up between the ministry and its subordinate R&D organisations. Ideally the *zakaz-nariad* should cover a major programme or project and not simply block funding for an institute. The fund is formed primarily on the basis of deductions from the profits of ministry enterprises. The size of the deductions are determined as a percentage of the ministry's 'normative net product' indicator (a sort of measure of value added) or sometimes planned sales revenue. The percentage is set for each ministry by Gosplan, GKNT, and the Ministry of Finance. Some input into the fund is also available from the extra profits realised by enterprises on the basis of price premiums attached to new high-quality products.[29] The stated intention at the time that EFRNT was introduced was to cut down on the amount of *melkotem'e* that had been characteristic of the heavy reliance on contracts under the old system.[30] However, the actual effect was steeply to reduce the amount of central budgetary allocations being made available to branch R&D. Thus between 1971 and 1980 the EFRNT percentage of R&D for industry as a whole rose from 10.4 to 47.2 per cent. The percentage from the state budget declined from 31.0 to 7.2 per cent.[31] The result of this was a substantial fall in the amount of fundamental research funded in the branch sector. When in 1969 EFRNT was introduced into the Ministry of Electrotechnical Industry 20 per cent of the ministry's science expenditure went on fundamental research. By 1972 it had already dropped to 15 per cent. The 20 per cent of the fund formally set aside for fundamental research since 1975 does not seem to have helped the situation. In one institute of the Ministry of Electrotechnical Industry the percentage of work done on exploratory projects fell from 8.2 per cent in 1973 to 7.5 per cent in 1975 and 2.7 per cent in 1977. The percentage of such work done in the machine-building ministries halved between about 1973 and 1983.[32] It also appears probable that the *zakaz-nariady* used to distribute EFRNT funds continued to be allocated on an institute rather than project basis.

EFRNT was a planning reform of rather narrow scope, being applicable only to the branch sector. But there was another development at about the same time that EFRNT made its appearance which was seen as being of much broader application. This was the use of the so-called 'programme-goal method' (*programmno-tselevye metody*, PTsM) in the planning process. PTsM was designed to attack not only the excessive *de facto* decentralisation of the traditional system, but also two other problems: firstly, institution- rather than project-based planning, the former giving the planning process a considerable degree of inflexibility since existing institutional structures were not easily diverted into new areas of concern; secondly, a lack of integration between research and production plans, creating serious problems of impractical research and low levels of acceptance and diffusion of research results by production organisations. The new approach was to overcome these problems: by ensuring that programmes were drawn up by high-level groups of specialists working closely with political decision-makers, with areas chosen for programme treatment receiving priority in terms of resources; by basing them on the achievement of goals rather than simply the operations of existing structures, and indeed it was the intention that most programmes would require the cooperation of a number of institutions; and by organising the programmes so that they covered the whole science–production cycle.

Thus the key elements of PTsM are a stress on the long-term nature of planning and concentration on the planning and funding of specific goal-oriented projects rather than institutions. The approach was developed gradually in a series of decrees and new procedures from 1966.[33] As the system now stands a 15-year plan is drawn up for the economy as a whole, containing and based upon a 'Complex programme of scientific and technical progress for the next twenty years.' On the basis of the 15-year plan and the 20-year programme the Five-Year Plan for the economy as a whole is drawn up, including a Five-Year Plan for Scientific and Technical Development, which in turn contains two sections – a section for scientific and technical problems, for which GKNT is responsible, and a section for *vnedrenie* under the control of Gosplan. Another important plan, but one which is not included in the state Five-Year Plan, is the five-year plan for natural and social science research, drawn up and supervised by the Academy of Sciences.

The adoption of the new approach at the national level led to its widespread application throughout the system, with programmes being set up between branches, including between the Academy and ministries, or within single republics and regions, branches, and even institutes. At institute level the programmes have entailed reorganisations along 'matrix' lines (see Chapter 6).

The programme-goal approach has been adopted on a wide scale. In the Tenth and Eleventh Five-Year Plans (1976–85) about 25 per cent of total scientific expenditure went to national S&T programmes. In the Eleventh Plan this represented 39 billion rubles, of which 11.5 billion were spent on running costs and 27.5 billion on capital expenditure. In the Twelfth Five-Year Plan it has been recommended that expenditure be trebled over the Eleventh.[34] In the early 1980s in a survey of Academy institutes an average of 50 per cent of budget funding went on programme work.[35] In Latvia in the mid-1980s 35 per cent of science expenditure went on programme tasks.[36]

While the approach has received an enormous amount of publicity and been introduced on a broad scale, there has been very considerable dissatisfaction with it. All the indications are that it has not been able to avoid the pitfalls of the traditional planning system. Thus the evidence is that programmes still tend to be simple compilations of research projects sent up from below – one source refers to up to 50 per cent of programmes being made up of projects coming under the 'others' category, i.e. projects that cannot be integrated into the ostensible goal of the programme.[37] Some Soviet commentators have suggested that programmes are often no more than the invention of institutions and individuals interested in the priority funding they might get for their own 'pet' projects.[38] It is further considered by many that the approach has been far too widely applied, because of the desire for priority resources among participants and the usual Soviet tendency towards overkill. This means that the approach is no longer a reliable indicator of priorities.

The 'compilative' nature of many programmes means that their goal orientation is often illusory, and this can be seen in the vagueness of the names and goals of so many of them. Goal orientation is further weakened by the fact that the programmes have not been able to win for themselves any practical independent existence outside the traditional planning system. While programmes do have separate lines in national plans, no work will be done on them until the tasks that make them up are included in the plans of individual operational units. Indeed, given that programmes are often set up in areas in which no existing institutions and hierarchies are interested, the programme tasks often fail to be included in regular plans or are given a low priority when they are. This particularly applies to the sections of the programmes dealing with the implementation of research results in production – despite the expectation that programmes would cover the whole research–production cycle it appears that they often do not. Either the implementation stage as a whole is not included in the programme or the relevant tasks are not included in production plans by Gosplan and the branch ministries.

Another interesting criticism of the project approach is that it gives too much power to scientists in the determination of resource allocations to major projects. Scientists, even if of the 'scientist manager' type, dominate the councils and commissions which draw up draft programmes, and some commentators consider that there are inadequate external controls (either bureaucratic or public) over their decisions. The effect is strengthened by the fact that programmes are usually run by the supplier rather than the customer. The most striking example of the perceived results (and of the ability to attack 'sacred cows' in the Gorbachevian Soviet Union) has been the recent savage press criticism of the Soviet magnetohydrodynamics programme, with a 500 million ruble project being launched on nothing more than the promise of results from the researchers of the Institute of High Temperatures.[39]

The programme-goal approach is still widely used. However, criticism continues and it receives far less publicity than it has in the past. It appears that the new MNTKs are an attempt at improving or supplementing the approach by reducing the number of areas covered, and so strengthening prioritisation, and by strengthening and increasing the institutionalisation of management structures.

It seems unlikely that any of the changes of the 1970s or early 1980s significantly changed the essential nature of the traditional system. The capacity for prioritisation that was already there probably was not increased. Indeed perhaps the increase in bureaucratisation implied in the greater degree of formalisation could have produced a weakening of prioritisation, as competing interests were able to take advantage of bureaucratic consultation procedures. The institution-based nature of the planning system and the separation of science and production were also still evident.

Decentralisation under Gorbachev

As in so many other areas Gorbachev has taken a radical approach to reform of the science planning system, although without adopting any entirely new concepts or approaches. The basis of his approach is 'full *khozraschet* (cost-accounting) and self-financing', and as such is intended to be identical with new decentralising procedures introduced into industrial production. Indeed research organisations, in all branches of the science sector, are now governed by the recent Law on the State Enterprise in the same way as industrial enterprises are. The meaning and intention of the new approach is summed up in the September 1987 CC-CofM decree 'On transferring scientific organisations to full *khozraschet* and self-financing':

> Proceed on the basis that scientific organisations like production enterprises are socialist commodity producers and their work must be based on the principles of full *khozraschet* and self-financing. They provide for their scientific–technical and social development with resources earned through the realisation of developments for users, and bear full responsibility for the results of their economic activity.[40]

In other words, they must earn enough money through satisfying the demands of users to cover costs and have enough left over for reinvestment and compensating staff. The decree also states that 'in the absence of orders for research and developments and in the case of the prolonged fruitless work of a scientific organisation and the failure of its superior organisations to achieve effective work the scientific organisation ceases to operate'.[41]

The basis of all the economic activity of research organisations is now to be the contract. This applies to both theoretical and applied research. However, for fundamental and other non-commercial forms of research, including major projects beyond the capacity of individual customers, the contract takes the form of the *goszakaz* (*gosudarstvennyi zakaz*, state order). The *goszakaz* is the by now infamous concept introduced into industrial management to replace the traditional plan task. It is in fact an obligatory task given to the operational body by bureaucratic superiors and has little of the sense of a mutually voluntary contract. What is supposed to be special about the *goszakaz* as distinct from the traditional obligatory plan task is that it covers only top-priority items and therefore should not take up the full capacity of the enterprise or institute. Any extra capacity can be used by the enterprise or institute as it desires.

For the science sector in particular the *goszakaz* system is seen as being importantly different because it entails the funding of major projects and not the block funding of institutes as was typical of the old system. Particularly with what is claimed will be the increasing use of competition (*konkursy*) between rival bids for a particular project it is hoped that the new system will lead to increasing severity being applied in the funding of research work and therefore of the institutes that do the work.

The *goszakazy* are financed either directly by GKNT, in the case of inter-branch projects, or by the Academy of Sciences, the State Committee for Education, or the relevant ministry. Ministries pay for their *goszakazy* from the new Fund for the Development of Production, Science and Technology (FRPNT). This fund replaces the EFRNT, and combines ministry money for funding production as well as R&D. It is formed in much the same way as the EFRNT, i.e. by deductions from the profits of enterprises.[42] It appears that the fund is likely to take about 50 per cent of enterprise profits.[43] As well as ministry *goszakazy* the FRPNT is to fund tasks undertaken within the framework of MNTKs and the Complex Programme for Scientific–Technical Progress of the Council for Mutual Economic Assistance.[44] According to the September 1987 decree the proportion of the fund going to such major tasks is determined by Gosplan and GKNT, while the tasks themselves are itemised and the relevant contracts are concluded by the ministries.[45]

That portion of the capacity, and costs, of institutes not covered by *goszakazy* is to be financed by direct contracts with user organisations, be they production enterprises or other R&D organisations. How much contract work the institute takes on is its own business, within the constraint of having to pay its own way. The September 1987 decree states that the price for contract work is agreed by the contracting parties. One recent commentator has stated that there are no plans to set prices centrally for contracted services.[46] However, another source, in describing the situation in the Ministry of Instrument Making, states that the ministry recommends a range of profitability for its research organisations and that institute and user negotiate within that range. The finally agreed price, relating to a particular level of profitability, is confirmed by the ministry.[47] As we shall see below, pricing is a sensitive issue at the moment.

Some information has begun to appear on the *khozraschet* system in operation. The indications are that it has been applied with some degree of ruthlessness, and therefore not surprisingly many institutes have been struggling financially.

Firstly, institutes seem not to be protected in the way that production enterprises are by ministries giving them *goszakazy* covering 100 per cent of their capacity. This practice is clearly against the spirit of the reform, but suits most managers who prefer to have fixed targets and most importantly the more-or-less guaranteed access to materials and supplies that the *goszakazy* carry with them. However, on the information currently available, 70 per cent seems to be about as high as institutes, particularly in the branch sector, can expect. This is the figure in the All-Union Institute of Cable Industry. The All-Union Research Institute of Genetics has been allocated 67 per cent, but warned that this figure will fall in future. For the chemical-timber sector the overall level is 63 per cent.[48] It is certainly being presented as if institutes are having to go out and find *goszakazy* and contracts in order to stay in business, although it should be mentioned that none of the closures of institutes reported in the last year or so have been linked to bankruptcy under the new *khozraschet* provisions. However, as 1 January 1988 approached, the date set for the introduction of the system into the branch sector, administrators were warning that chaos was imminent, with at that stage, for example, only 20 per cent of institutes in the machine-tool sector having sufficient contracts to pay their way.[49] It was said that at the beginning of 1988 branch institutes as a whole had found orders and contracts to cover only half their costs. Some have had trouble meeting their wage bills, and amortisation is also seen as posing a major financial burden.[50] Because of the traditional lack of operating funds in Soviet research institutes they have had no reserves to tide them over the initial difficult period of finding orders and waiting for the first returns to come in.[51]

The supply problem is also already becoming evident. Contract research is not tied to any planned supply of equipment or materials, meaning that such things have to be obtained by 'roundabout methods', something which it is said produces demoralisation in institutes.[52] It is considered that this difficulty will be overcome by the introduction of 'wholesale' trade, i.e. the unplanned distribution of goods. The science sector was in fact supposed to be the first to transfer to wholesale trade from the beginning of 1987 (although wholesale trade here seems to have meant simply that institutes approached local supply agencies in their own time for small-scale scientific equipment and supplies, but that local supply agencies still obtained their stocks through planned allocations made by central agencies).[53] Science managers still talk of setting up a wholesale system,[54] but one suspects that many fear that supplies will become even more uncertain, and probably more expensive, under such a system.

A number of measures have been adopted to help institutes in these difficult circumstances. Firstly, they have been given greater access to credit, with bank credit indeed now being seen as an important component of the long-term financial management of Soviet science. Bank credits are also used to overcome short-term difficulties faced by organisations unable to meet their commitments, although in some cases the banks demand evidence – in the form of signed contracts – that future receipts can be expected.[55]

The ministries have also helped out, on occasions by providing funds on a clearly short-term basis.[56] However, there is also evidence that they are trying to 'subsidise' weaker institutes by spreading around *goszakazy*, ensuring that everyone gets at least some guaranteed income.[57] As already mentioned, this approach flies in the face of the aims of the reform.

In some articles one gets a sense of near panic among institute managements as a system which they barely understand but which clearly seriously threatens their positions is being introduced with at least some degree of ruthlessness. However, some commentators report with relief that institutes which have been operating on the system for a while have come to understand and come to terms with it, and that beneficial changes have been seen as a result.

Perhaps a good example is the All-Union Research Institute for Genetics. It is an institute which does fundamental research but can also produce commercial research and products. It has received 67 per cent *goszakazy*, but has managed, not without difficulty, to find contracts to cover the rest of its costs. These contracts have been a mixture of major projects and minor money-spinners. The institute has had to reorganise itself in order to make it possible to satisfy the demands of its major customers. One change has been the introduction of *khozraschet* between the units of the institute. Money has also now become available for the hiring of outside specialists on a consultancy basis. The institute has had to devote a lot more attention to providing training, service, and consultancy services for its customers, both to keep them happy and also to bring in extra income. The institute has become more aware of the need to control overhead costs and it has realised substantial savings in its transport and heating costs. If all outstanding contracts are fulfilled successfully the institute staff stand to gain income increases of on average 10–15 per cent, this money in fact being given to about 20 per cent of the staff who will receive increases of up to 40 per cent. The management of the institute does not hide the fears and difficulties it faces in coming to terms with the new system. However, the tone is now of an institute which is more flexible and responsive than in the past. This sounds like the response that the reformers have been looking for.[58]

Certainly if an institute has something to sell it can do very well under the new conditions. The most spectacular example is the Institute of Microeye Surgery, run by Sviatoslav Fedorov. It has developed very advanced 'production line' techniques in eye surgery which allow it to do 400 operations a day. This has provided it with an enormous income, and it is now in the process of building an empire of monumental proportions. The staff of the institute are also receiving far above average earnings, including that gained through the buying of shares in the enterprise.[59]

For many institutes such a happy situation is made possible by their monopoly position and their willingness to charge highly inflationary prices. The Soviet R&D sector has grown up deliberately based on monopolies, and it seems very unlikely that the new stress on competitive bidding and parallel projects will be enough to overcome this institutionalised monopolisation. It is noteworthy that in the September 1987 decree, while one paragraph talks about the need for competition, another establishes that the prime function of the ministries is to provide for a 'united scientific–technical policy'.[60] The ministries are unlikely to take that as other than a call for the maintenance of monopolies.

The other problem is inflation. Clearly this is going to be a general problem for Gorbachev – if you introduce true market mechanisms into a deficit economy inflation has to be the result. The signs are certainly there in R&D. One science economist has estimated that S&T costs will increase by a factor of between one and a half and two times in the current five-year plan. A GKNT department head has complained that in cases where the customer must have the R&D work in order to fulfil its own *goszakaz* and the supplier is in a monopoly position, price increases of 40 to 60 times for R&D are not unknown.[61]

The Academy of Sciences and the *vuzy* are particularly nervous about the new system, since it seems seriously to threaten their access to funds for non-commercial research. Academy leaders have stressed that it is not possible to apply *khozraschet* to fundamental science, although some have expressed a willingness to accept the *goszakaz* approach.[62] This seems to be what has happened, with most scientists from these sectors seeming to assume that the new system will mean that their work will continue in fact to be financed from the state budget.

The implications of the system are that many institutes will go out of existence. It is not yet clear that the authorities have entirely accepted the consequences, either in the science sector or in the economy as a whole, of widespread bankruptcies and unemployment. One commentator has pointed ominously to the results of *khozraschet* in Polish institutes, where scientific employment fell by half, the most mobile and dynamic staff being those to leave, and the proportion of national income devoted to science decreased.[63] But probably the main problem, as it so often is when the talk turns to economic reform in the Soviet Union, is that there are still not strong enough incentives for industrial managers to welcome innovation. Some commentators have noted that there is no evidence of recent economic reforms leading to an increase in interest in innovation; some have even noted a decrease, as enterprises find themselves short of funds.[64] The consequence of such a situation is that there is no guarantee that even good R&D will survive. There are already signs of further falls in the amount of fundamental and exploratory research being funded.[65] This is presumably one reason for the continued place for centralised funding of not only fundamental research but also 'important' applied research, through GKNT and central ministry funds. We shall have to wait and see whether this part of the R&D funding process will come to dominate, and whether in the traditional way the ministries and their head institutes will reestablish their monopolies and institute directors will regain control of their block funding.

Conclusion

The Soviet science planning system still appears to face its traditional and perhaps inherent problems. If it is a truly rigid centralised system some successes are gained through prioritisation. But for the system as a whole irrationalities of resource distribution spring up and perhaps even more importantly incentive and responsibility are severely reduced at operational level. Inevitably a considerable degree of *de facto* decentralisation becomes evident, with central plans becoming little more than confirmation of the current activities, or lack of activities, of operational units. Often the response to these problems is to tighten up the central planning system further, as occurred in the 1970s. This tends to compound the problems. Occasionally, as at present, a 'marketising' reformer comes along and tries to formalise decentralisation as a direct attack on the incentive and responsibility problems. This leads to some loss of central control of priorities, but more importantly raises the question of what to do with institutes and researchers that cannot pay their way. That is a question with enormous political and social implications. The response of the current regime to that question, within the context of the Soviet system as a whole, is not yet clear and probably has not been determined. But there are many signs of a desire, or need, to leave a way out, a way out which can only be back in the direction of centralisation.

Chapter eight

Research managers

The basic principle of Soviet management is *edinonachalie*, which is usually translated as 'one-man management' and refers to the total decision-making power and responsibility possessed by a single manager in any working body. This makes the director, the *edinonachal'nik* of the research institute, the bearer of ultimate power and responsibility.

That authority also devolves down to the subunit heads in the institute – the sector, department, and laboratory heads. Thus while the director is the key figure, laboratories headed by high-profile scientists can operate as mini-institutes, or feudal fiefdoms to use the phrase of the physicist Lev Artsimovich.[1] Therefore many of the comments to be made here about the director could also be applied to lower-level managers.

The director is expected to be both scientific and administrative head of the institute – as we shall see shortly this is a key feature of Soviet science management. As scientific head he determines the type of research to be done in the institute. Usually the director's general field and approach are formally laid down as the 'major research directions' (*napravleniia*) of the institute when it is founded. Thus the new Institute of Mathematics of the Bashkir branch of the Urals Department of the Academy of Sciences has been set up to work within the traditions and aims of the Ufa school of complex variable functions, founded by A. F. Leont'ev. The founding director of the institute is Leont'ev's student, V.V. Napalkov.[2] A new director coming into an established institute is able to change the type of research done in the institute radically, including bringing staff with him from his old institute.[3]

Given that he bears administrative responsibility along with his scientific leadership, he has considerable *de facto* power to distribute funds within the institute despite the apparent rigidities of the central planning system while the lack of access of researchers to personal funding means that they have no choice but to rely on the funds at the disposal of the director. One of the most important and most difficult responsibilities of the director, and the managers below him, is to obtain the resources needed to push the research programme ahead. This is not simply a matter of obtaining funding but, perhaps more difficult, of organising the construction and equipping of facilities. The vagaries of the Soviet supply network ensure that this is perhaps his most important and difficult task. It requires great 'pushiness' (*probivnye kachestva*) and good scientific and bureaucratic contacts. We have seen some examples of the activities of research managers in these areas in Chapter 7.

Management style

We have just mentioned the combination of scientific and administrative responsibilities in the single person of the *edinonachal'nik*. This is the first of two qualities that research managers are expected to possess within the official 'culture of science'. They are expected to maintain a research commitment and reputation, rather than openly devoting themselves entirely to administration. The second quality is that they are expected to exercise the correct 'democratic' style of management. While they are very much in charge, they must exercise authority in neither an authoritarian nor a 'liberal' way.

Science versus administration

Let us now deal with the first of these requirements. An important issue in research management throughout the world is what the best mix of research and administrative responsibilities is for senior managers, and how that mix should be arrived at. It is also an issue of debate in the Soviet Union. Some Soviet sources make much of and probably exaggerate the fact that, unlike in the West, research management positions in the Soviet Union are held by active scientists. There is not the differentiation between researchers and managers within the scientific community, and progress along the administrative ladder presupposes and is indistinguishable from success in the research sphere. It is recognised that there are problems in such an approach, primarily that talented scientists waste valuable time on administrative duties and that they might not be very good at administration. However, the advantages are seen as outweighing the problems. The advantages are that administrative control is simplified since managers and managed speak the same language and share the same interests, while a scientifically competent manager is more likely to retain his own self-confidence and the confidence of his subordinates. In general the Soviet approach is seen as working to keep to a minimum the bureaucratisation of science and making possible a more relaxed and flexible management style.[4] It is certainly in the interests of the scientists to stress the importance of having scientists running science, especially in the context of the Soviet tradition of heavy-handed bureaucratic interference in all areas of society.

As one would expect, an institute director usually has some research background. This seems particularly to be the case in Academy institutes. Virtually all directors have some background in research, and indeed most can be rated as major figures in their fields. In 1961 of 117 directors of central Academy institutes 36 were academicians, 42 corresponding members, 28 doctors of science, and 11 candidates of science.[5] In 1962 83 per cent of the directors of the institutes of the Ukrainian Academy of Sciences were full or candidate members of the central or Ukrainian Academies. All directors had at least a candidate's degree. Of 50 directors in the same Academy in 1967 21 were full academicians, 17 corresponding members, 11 had doctorates, and one had a candidate's degree.[6] In the Siberian Academy in the 1970s 93.2 per cent of directors had doctorates and the rest had candidate degrees.[7] Many try, if we can believe published reports, to maintain at least some presence in the laboratory despite heavy administrative workloads. They retain for themselves the formal headship of a laboratory or maintain apparatus on which to do their own individual research.[8] However, many commentators accept claims that they in fact do any genuine research with a large grain of salt. Even Soviet commentaries admit that many directors would not have entered a laboratory in decades.[9] But a research reputation is seen as important for maintaining their prestige among both their peers and perhaps also their political superiors. It is stated over and over again in Soviet sources that research managers, particularly in the Academy, should be above all scientists. An Academy researcher wanting promotion must work hard therefore to maintain at least the appearance of research activity and success.

In branch institutes the situation is rather different. Again there are directors of branch institutes who are or have been genuine scientists. However, there are many, and there always have been many, who lay no claims whatsoever to a scientific background.[10] Sometimes directors have a production background. On occasions this practice is reported as having a beneficial effect, particularly in institutes where management has been a problem. Presumably it is considered that a firm hand is needed and cannot be found within the scientific community.[11] Despite this the practice is generally seen as an undesirable one.[12] In recent times it has become possible to criticise the appointment of 'professional managers' with party apparatus backgrounds as directors.[13]

Clearly the issue of what type of people are institute directors is not as straightforward as those claiming the existence of the classic 'scientist manager' would have us believe, although a great deal more empirical work needs to be done before we can come to any firm conclusions.

We must now ask ourselves whether the Soviet 'scientist manager' approach is a good thing or not. One cannot say that the alternative Western approach of career streaming is necessarily preferable, since it is in fact recognised as one of the most problematic issues in Western research management. But equally the Soviets are not entirely satisfied with their approach. Firstly, it is clearly considered that the quality of management provided by untrained scientists is a serious problem. Either they leave it entirely to the administrative personnel in the institute and superior bodies, thus in effect leaving the management of the scientific process in the hands of non-scientist bureaucrats, or else they manage very poorly. The greatest source of staff dissatisfaction with managers is incompetence in obtaining equipment and materials.[14]

Secondly, and perhaps more importantly, it is claimed that scientist managers are in fact being turned into bureaucrats but without anyone openly acknowledging the fact or admitting the problems that this produces. The pressure on managers to look scientifically productive while they are in fact swamped by administrative tasks increases the temptation to engage in unethical behaviour.[15] At the very simplest the need to keep up a publication rate leads to plagiarism. Reports of classic plagiarism, i.e. someone presenting work as his own which was in fact done by someone else and without the knowledge of the real author, are relatively rare, and there seems no reason to believe that it is more prevalent than in the West.[16] Far more attention is devoted in Soviet sources to that other phenomenon which is by no means unknown in the West – research managers putting their names to the work of subordinates, with the knowledge of those subordinates. (Sometimes such plagiarism is done on a commercial basis, with in particular theses being written for payment.) It is a sensitive issue in the Soviet Union, with the same arguments for and against as in the West. Certainly young researchers are often glad of the extra notice their work will attract if their superiors' names are attached to publications. But overall one gets the impression that not only is such expropriation of research results endemic, but it is often carried out in a dictatorial and immoral way. A 1984 article seems to be speaking of a common situation in claiming that if a rank-and-file researcher wants to obtain funding and other resources to work on a project he or she must include superiors as co-authors of any articles or discoveries that result. The more important the work the larger the number of co-authors that are likely to appear.[17] Another article refers to researchers who refuse to take managers on as co-authors being forced out of institutes.[18]

But the research manager's concern with his scientific reputation goes far further than the need to keep up his publication rate. He also feels the need to protect those theories and approaches he has championed throughout his career. This means essentially allowing no opposition or external threat to his 'school', i.e. the group of students and colleagues he has built up around him. The school is usually equivalent to the institute, but if at all possible the manager will extend his domination to the whole discipline, through control of the editorial boards of journals and the major councils and commissions that determine research priorities and examine dissertations. The result is a strong element of 'monopolisation', usually ruthlessly enforced. We shall look at some concrete examples of the type of behaviour involved later in this chapter. The consequences are a conservatism in Soviet science, as well as a demoralisation of the victims or potential victims of the 'monopolies'.

'Democratic' style

The second major quality officially expected of Soviet science managers is, in Soviet parlance, the 'democratic' style. This essentially means that the authority figure undertakes to exercise authority in a way that takes account of the desires, or more accurately, the objective interests of his or her subordinates. By definition, in Soviet conditions these interests are ultimately identical with the interests of the state and society as a whole. Consultation with the workforce is theoretically part of the process, but there is by no means any guaranteed right of the workforce to have its opinions acted upon. The authority figure is directly responsible for the taking of decisions, and firm discipline and tight work practices are expected. In one source the democratic style is defined as that under which 'persuasion of subordinates is combined with demandingness towards them'.[19]

The democratic style is contrasted with the extremes on either side of it, the 'authoritarian' style and the 'liberal' style. The authoritarian style entails petty and arbitrary control by management, with no regard taken of the legitimate interests of society or the workforce. The liberal style ('liberal' is a word with particularly negative connotations in the Soviet political and management lexicon) involves an unwillingness to enforce strict work practices and probably more importantly an excessive concern for respecting the wishes of the workforce, including of course a willingness to tolerate 'anti-Soviet' behaviour.

Perhaps the best sense of the democratic style can be gained from the biographies of Soviet scientists. These tend to written in a formulaic, indeed often hagiographic, way and a number of positive characteristics of their heroes appear in virtually all of them. On the one hand, these include courtesy and the equal treatment of all regardless of rank, including the inevitable doorman or cleaner with whom our hero is on first-name terms (often the only person with whom our hero is so familiar, since usually he is very correct and formal in his relations with his colleagues). He is always prepared to help his colleagues with their personal problems, even to the extent of engaging in some minor *blat* (informal influence), for example getting someone a flat or getting access to a top doctor for a colleague's sick wife. He takes a close and personal interest in the work of all his staff and students, being helped in this by his phenomenal memory, but does not like to deprive them of their independence or initiative. He accepts with good grace, and even pleasure, those very rare occasions when his scientific judgement is proved incorrect by a subordinate who has persevered with an experiment of which he disapproved. He is invariably a workaholic, maintaining an extraordinary research, teaching, writing, and administrative work load, and often works himself to death at a relatively early age. Very often he is editing some work in his usual precise and methodical way on his death bed.

On the other hand, our hero is always very 'demanding' (*trebovatel'nyi*). This is best demonstrated by his refusal to maintain relations with anyone who shows anything less than a total commitment to science. He is impatient of those who give pressure of work as an excuse for not completing an experiment on time, of those who present underprepared papers at seminars, and of those who find lack of knowledge of foreign languages an obstacle to reading the foreign literature (our hero is usually fluent in at least six languages).

I do not wish by my ironic tone here to suggest that I consider everything in these biographies to be lies. The Soviet Union's leading scientists, like specially talented people in any society, probably are workaholics, while those coming from old intelligentsia families in particular are well educated and erudite. They are probably as tactful or tactless, courteous or rude, as any other group of people. But that they are all so similar in character seems unlikely, while many accounts we have of less famous science managers suggest that the demandingness often takes precedence over the tact and courtesy. Given the behaviour to be discussed below, it is particularly difficult to believe that 78.5 per cent of research managers are democratic (as against 10.9 per cent who are liberal and 10.6 per cent who are autocratic), as one survey apparently revealed.[20]

The two best examples of the correct management style are usually considered to be Igor' Kurchatov and Sergei Korolev, the heads of the bomb and space programmes respectively, because they had the right combination of flexibility and toughness.[21] There is no doubting the success of these programmes, and no reason to doubt that the two programme heads operated in the way claimed – while programme staff worked according to rigid discipline and tight deadlines, the programme heads consulted closely with their subordinates and could well defer to their judgement on how particular tasks should be undertaken. It is in fact an approach which has proven its worth in other systems as well, ranging from similar US defence programmes – the Manhattan Project and the Polaris programme, for example – to Florey's management of his penicillin programme.

However, when one reads accounts of the management styles of Kurchatov and Korolev one realises how narrow the gap is between management success and disaster. The fact that their style is described by Soviet commentators as democratic shows how close to authoritarianism Soviet democracy is. They were prepared to grant their subordinates independence within the parameters set by them. But anyone trying to operate outside or to change those parameters would find themselves in trouble. In the programmes in which Kurchatov and Korolev were working this approach caused few problems – they had the authority derived from the importance and urgency of their tasks and the support of their political superiors to enable them to get away with it.

But it seems unlikely that such an approach would be successful in the bulk of scientific circumstances, either in the fundamental or the applied sectors. This has even been publicly recognised by academician K.S. Frolov, the Academy's 'chief engineer', who recently directly referred to the Kurchatov–Korolev style of centralised authority as being no longer appropriate (although he did also comment that Kurchatov and Korolev were not as authoritarian as people make out).[22] Numerous other commentators, without going so far as to disown Kurchatov and Korolev publicly, make it clear that one of the major problems facing Soviet science is the excessively authoritarian nature of its managers.

It could be said that the dangers of authoritarianism are inherent in the officially encouraged style of management – managers are under great pressure to give the appearance of continuing research success, and they have the opportunity provided through the bureaucratic powers granted the 'one-man manager' to ensure that such an appearance is maintained. These pressures are seen by many outside observers as being reinforced by the politico-ideological aspects of bureaucratic control in the Soviet context.[23]

Authoritarian managers

To what extent then are these authoritarian tendencies and behaviour evident in Soviet science? Certainly the Soviet press contains a regular diet of horror stories. The incidence of such stories has increased recently, in the context of a campaign against corruption which has been applied to the science sector no less than elsewhere. However, such stories could also be found well before Gorbachev came to power. Sometimes it is simply a matter of senior research managers expecting a subservient attitude from their subordinates. Academician Moiseev reports with distaste the boastful comment of an institute director hosting a barbecue at his country cottage: 'Now I have doctors of science cooking my *shashlyk*.'[24] But often the authoritarianism takes on a more sinister form. The victims are usually those who are pushing a scientific development which comes into conflict with the favoured views of the director, those who have tried to do something about the director's misdeeds, or simply those who are seen as 'difficult' characters.

Perhaps the most diplomatic approach on the part of the director is to ensure that those potentially falling within these categories are not allowed into the institute in the beginning. V.V. Belousov, the best-known opponent of plate tectonics in the Soviet Union, when asked whether there were any supporters of plate tectonics in his Institute of Earth Physics, replied:

> There are none. It doesn't mean that I exert pressure to bear on my colleagues, but I categorically demand a fair treatment of facts from them. And given such an approach I think that there will never be any supporters of the new global tectonics theory at our institute.[25]

The journalist reporting this goes on to say that she knows that there are in fact people in the institute with positive attitudes towards plate tectonics, but that they prefer to keep quiet about it.[26]

About the mildest measure that can be taken against those out of favour is their exclusion from access to apparatus. Timothy Toohig reports from the accelerator at Dubna, without drawing any authoritarian conclusions, that the closing of the joint experiment which Fermilab had been doing there because of the refusal of institute management to give more beam time could have been the result of jealousy between research groups. Such jealousy, he states, is especially strong at Soviet accelerators, 'where resident groups become entrenched on the experimental floor'.[27]

Powerful managers not only control the apparatus; they also control the journals in which results can be presented. The membership of editorial boards is an important issue in Soviet science, and recent complaints about the misuse of the anonymous refereeing of submitted manuscripts suggest that control of editorial boards is often abused.[28] The requirement that any manuscripts, whether for a journal or a conference, be checked by the institute makes it very difficult for researchers to find alternatives to the 'monopolised' journals.[29]

The blocking of dissertations is also a tried-and-true method. In one case the deputy director of an unnamed institute tried to block the dissertation of a researcher whose work he did not like on the grounds that he was politically unreliable, as demonstrated by his unwillingness to serve in the citizens' militia (*druzhina*).[30]

We also have cases of replications being rigged to discredit the research of rivals. One example concerns a senior research worker at the Sverdlovsk Research Institute for Chemical Machine Building who failed to pass an efficiency review board. It was claimed in the press that he was hounded because he had developed a better system than his department head and on his own initiative had had it tested in production conditions. It was stated that comparative trials of the two systems had then been rigged to discredit his work, while one of his supporters in the institute, the head of the Patent Department, was dismissed for 'inadequate qualifications'.[31] Disapproval of a research project and its findings can often lead to the destruction of equipment and data.[32]

Some of the cases here have involved the dismissal of recalcitrant researchers. The efficiency review (*attestatsiia*) system which is periodically applied in Soviet science as a way of introducing some turnover in personnel is a boon for authoritarian managers, since they are usually able to organise the system in such a way that it is their opponents who are found wanting. It is openly admitted that the most recent efficiency review campaign, introduced in 1986, has been disastrous in this regard.[33] Once dismissed it is sometimes arranged so that other work cannot be found.[34]

One might have thought that dismissal, long-term unemployment, and the destruction of years of work would be punishment enough. But some stories go further. Articles appearing in the newspaper *Sotsialisticheskaia industriia* in 1987 referred to the situation in the Azerbaijan Academy of Sciences in which all senior posts are reserved for people from two small geographical regions. Any 'outsiders' trying to break this monopoly are subjected to extreme victimisation, including sabotage of their research, dismissal from their jobs, and even imprisonment on trumped-up charges.[35]

Another echo of the Stalin era is the number of cases which have been reported recently in which political and ideological denunciations have featured. We have already mentioned the victimisation of the researcher who refused to serve in the *druzhina*. In the Central Town Development Institute charges of political unreliability were made against staff who dared criticise the director and party secretary. A researcher in a branch institute in Tula, who had made himself unpopular through his single-minded commitment to true scientific endeavour, found himself convicted of slandering the Soviet state after being denounced for declaring in the tram on the way to work that his discoveries, which had not been adopted in the Soviet Union, would have been adopted in the West.[36]

A major recent example reported in *Izvestiia* was the removal from teaching duties and eventual dismissal (since reversed) from Leningrad State University of the noted specialist in upper-atmosphere aerodynamics, R.G. Barantsev. His troubles began when the head of the university's Chair of Dialectical Materialism determined that an informal seminar for young students and researchers organised by Barantsev on 'semiodynamics', 'the study in an interdisciplinary way of information transfer in complex systems', was ideologically harmful and an attempt to develop a new metatheory to replace dialectical materialism. It was also hinted in the article that Barantsev had earned the personal antipathy of his colleagues while he had been head of the Chair of Mathematics and Mechanics and that they might have set him up.[37]

The impression one gets from all these reports is a depressing one. But how widespread and serious are such phenomena? Could the reports we read be isolated cases, examples of journalists being taken in by cranks with persecution complexes, or politically motivated campaigns intended to shock? While admitting the need to exercise due care, I believe that the evidence suggests that a problem of significant proportions exists.

'Liberal' managers

There are of course 'good' Soviet research managers who do everything in their power to avoid authoritarian behaviour. Raisa Berg, it is true, includes in her category of directors who are truly committed to scientific truth above all else only one person, the biologist Astaurov.[38] Nevertheless one can find other cases of senior scientists standing up for colleagues in political difficulties. This was particularly evident at the height of the dissident movement in the mid-1960s (a time, it has to be said, when low-level dissidence was not a particularly dangerous activity).[39] But cases can be found throughout Soviet history. At the height of the Stalinist purges senior scientists were willing to intercede on behalf of arrested colleagues. For example, Sergei Vavilov and V.A. Fok complained, without result, to the procurator about the 1937 arrest of their colleague M.L. Bronshtein.[40] It is generally considered that it was the insistence of Petr Kapitsa that brought about the release from prison of Lev Landau in the same year.[41] Iurii Orlov has recently described the reaction of A.I. Alikhanov, the director of the Institute of Theoretical and Experimental Physics, when Orlov was dismissed from the institute and expelled from the party in 1956 for demanding that de-Stalinisation be taken further. Alikhanov complained to Khrushchev, who replied that Orlov was lucky to have been no more than dismissed, since some party leaders wanted him shot. Alikhanov then got Orlov a job working for his brother in Erevan. When in the post-Khrushchev thaw Orlov was invited to rejoin the party Alikhanov advised him not to, telling him 'not to give in to this shit'.[42]

In adopting this position scientists seem to feel that they are acting within the tradition of the Russian intelligentsia – the tradition that arose in the nineteenth century of maintaining a moral conscience in opposition to the government. The family trees of the Soviet scientific élite show a considerable degree of continuity from pre-revolutionary days, while memoirists often refer to the intelligentsia tradition. For example, in a recent biographical article on the mathematician A.A. Lyapunov, N.N. Vorontsov says of the whole Lyapunov dynasty:

> The characteristic traits of the Russian intelligentsia – humanism, breadth of spiritual enquiry, the principle of service to society – all this clearly brought distant and close relatives together, with the traditions being handed down from generation to generation.

He then gives details of Lyapunov's involvement in protests to the government about the treatment of genetics in the 1950s.[43] E.L. Feinberg, in writing of the special qualities of E.A. Tamm (who was from a provincial intelligentsia family), says:

> It seems that it was this form of the concept of civilised behaviour which was developed with special strength in a particular environment and in a particular time, namely in the best layers of the employed intelligentsia in Russia at the turn of the century, and it has come to us from then.[44]

Many might consider the Russian intelligentsia's 'social conscience' to be little more than a sham, all too easily compromised by the hard knocks of daily life. Nevertheless it is worth keeping an eye open for evidence of the existence of a self-conscious scientific élite with an 'alternative ideology'.

One even has to be careful in one's characterisation of party-beholden 'careerist' managers. When people like Mark Popovsky write about 'party hacks' at the top levels of Soviet science, they write as if they can by definition bring nothing but harm. However, a commitment to the party does not necessarily exclude a concern for science, even if only for purely careerist reasons. If their careers depend on the success of 'their' science, one can expect to find managers displaying skilled advocacy and protection of their institutes and the science sector in general. Clearly this entails at the very least a willingness to fight for funding and allocation of resources to the science sector – this is the normal behaviour of ambitious bureaucrats. But career-minded science managers often find they have to go a little further. To provide the appearance of science running successfully under their control, including their own research contribution, they rely to some extent on the goodwill of the scientific community – to accept their 'ghosted' articles and books, to vote in secret ballot for their membership of the Academy, etc. To obtain this cooperation some concessions have to be made – allowing 'unreliable' researchers to continue to work in 'unreliable' disciplines, giving scientists greater freedom in their research, and of course looking after them in terms of funding and apparatus.[45]

One should not get too carried away about the 'pro-science' behaviour of career-minded managers – they will submit totally to political pressure when required, and even when they are fighting for science they are fighting for 'their' science. Such activity is in no way inconsistent with – indeed is an essential part of – the 'monopolisation' of science referred to above. And one gets the impression that overall the damage done to Soviet science by the domination of authoritarian managers far outweighs the benefits that they might bring. The stories related in the press are shocking, and many Soviet commentators suggest that they are only the tip of the iceberg.[46] Western scientists regularly comment on the conservatism and rigidity of Soviet science that results from its domination by elderly autocrats, while Soviet commentators draw attention to the catastrophic effects on the morale, morality, and productivity of researchers (see Chapter 9).

A new approach to management?

The problems of Soviet research management have long been recognised in the Soviet Union, and are getting special attention in Gorbachev's reform programme. Some of the difficulties of the classical Soviet 'scientist manager' approach are recognised, and proposals for changes are put forward. The research management specialists, the *naukovedy*, call for the use of specially trained science administrators, operating according to all the best management principles.[47] This has been one of the grounds for friction between the scientists and the *naukovedy*,[48] since the response from the scientists is, hardly surprisingly, to oppose the idea of professional managers being put in charge of them. They have suggested a number of possible compromise solutions. Many scientists make the point that although it is essential that a scientist be in charge of a research collective it is not essential that this scientist be a leading scientist or the most talented in the particular collective. It is only necessary that his colleagues have confidence in his scientific knowledge and awareness. The engineer I.I. Artobolevsky is said to have considered that such qualities as 'knowledge of one's business, the ability to understand people and to predict their future creative successes' were no less important for a modern research manager than creative talent. He tried to push people with these qualities up the career hierarchy, and although his choices might have appeared strange at first they usually proved to be well made.[49] Many would say of course that these are the thinly disguised professional bureaucrats with scientific pretensions that have been causing all the problems.

Other approaches to the problem are that scientists should be provided with specific training in management techniques or that business managers should be hired, although in a subordinate position to the scientist director.[50] With regard to the first approach, it appears that there is no management training for scientists in Academy institutes, but that middle management in the branch sector do go through 'raising of qualification' courses which usually include management and economics training.[51] The Ukrainian Academy of Sciences is in the process of setting up an institute for 'raising the qualifications' of its managers.[52] But one could be sceptical about the value of these courses, not least because of the lack of people trained to teach them.[53] Further, one cannot imagine scientists being too pleased with the idea of management training. The disdain with which they regard *naukovedy* gives a clue to the attitude they would adopt towards specialist management training.

The second approach to improving management is to hire business managers. This in fact describes to some extent the existing situation. Most institutes have a deputy director for administrative affairs and the scientific secretary, who as we have already seen has substantial administrative responsibilities (see Chapter 6). A number of commentators feel that a new post of high prestige and power, and occupied by professional managers with no interest in or claims to a scientific career, should be introduced.[54]

Another approach suggested by some is to break institutes up into smaller units, to make their administration more manageable. There is a general discussion at the moment of whether Soviet institutes are too large, and the management issue is a key aspect of it.[55]

Finally some think the key is to reduce the emphasis on obtaining a directorship as a key stage in an academic career. The feeling is that directorships should be circulated more than at present, so that an unsuccessful or even successful administrator can resign his post without stigma.[56] This will allow top scientists more time to do creative research work, and by being more 'democratic' will improve relations within institutes.[57] It will also break down to some extent the link between scientific leadership and administrative power, and thereby reduce the dangers of 'monopolisation'. The retirement of O.I. Sumbaev as director of the Leningrad Institute of Nuclear Physics at the age of 55, in order to get back to creative work, is held up as an example.[58]

Under Gorbachev many managers are being given no choice in the matter. After a slow start the turnover of science managers has increased rapidly in the last 12 months or so. Regulations have been introduced into the Academy and *vuzy* that no one can hold a management post (from laboratory head up) beyond the age of 65 (full and corresponding members of the Academy are allowed to retain their management posts up to the age of 70).[59] After some initial resistance,[60] it now appears that the new regulations are being enforced, in the Academy at least. In mid-1988 it was announced that six academician–secretaries and ten members of the presidium had just retired. Although 20 over-age institute directors, three deputy directors and 160 subunit heads still remained in the Academy, the announcement suggested that their days were numbered.[61]

The purge has been imposed on science managers from above. Radical reformers suspect that such a purge might be no more than a one-off event, and suggest that the only way of reliably improving management is to break the power of authoritarian managers permanently by decentralising authority to institute workforces.

The first demand is for the election of institute directors and middle management. Under Gorbachev quite a lot of attention has been devoted to elections in all sectors of society, with the election of enterprise directors and foremen by the workforce apparently becoming reasonably widespread. It is a practice which is also appearing in the Academy, although one with which Marchuk and other senior Academy personnel quite openly disagree. Marchuk created an uproar when he responded to a question in an interview in *Izvestiia* on the possibility of the election of managers as follows:

> If we give the right to elect the director to the collective itself this could lead to negative consequences. They could well elect a good person, but there is no guarantee that he would also be a good scientist, capable of being the scientific leader of the institute . . .. The directors of institutes must be elected at the General Meetings of the Departments [of the Academy], which know well the level of work in the field at home and abroad. They can evaluate the potential of the candidates. We cannot adopt a 'laissez-faire' attitude here. The mood of the collective at the time could conflict with the interests of the state.[62]

Another senior academician put things more bluntly when he declared that the majority of staff of institutes are 'good-for-nothings' (*bezdel' niki*) and cannot be trusted to make a responsible choice.[63]

Marchuk's answer produced an apoplectic reply in *Literaturnaia gazeta* from a large group of Academy social scientists. The newspaper editorially supported the social scientists.[64] A number of Academy institutes went ahead and held elections for directors. In the Institute of Oceanology the incumbent director, A.S. Monin, got two votes out of 2,109; in four institutes of the Department of History the 'official' candidate was defeated in each case. However, it is clear that the management organs of the Academy still claim the right to overturn the popular choice. In the four history institutes the Department of History overturned or failed to confirm the election in each case, according to an *émigré* source.[65] Since then formal arrangements for the Academy have been published. These demonstrate an attempt at compromise. The procedure now is that it is possible and even desirable that a number of candidates be put forward to a vacant directorship. The choice is made by a secret ballot of the Academy members of the Department supervising the institute. Candidates can be proposed by the bureau of the Department, individual members of the Academy, or the Academic Council and 'social organisations' (i.e. party, komsomol, or trade union) of the institute. The opinion of the institute staff, determined at a general staff meeting or delegate conference in a large collective, is also to be conveyed to the election meeting. The deputy director for science, as before, is recommended by the director for confirmation by the bureau of the Department. But now his candidacy will be examined and approved by the institute's Academic Council. As for middle management, they are elected on a competitive basis by the Academic Council for a five-year term from a number of candidates put forward by the staff of the relevant unit. The staff votes in secret ballot to determine its preferred candidates. Provision is made for the appointment of an acting head by the director for no more than five years if no election takes place (it is not stated what circumstances might make this necessary). Also members of the Academy are not subject to election, but are appointed by the director.[66]

One doubts that the radical reformers are satisfied with this, and it is noticeable that these changes are contained in a new Temporary Statute for Academy institutes. Clearly there is room for further changes and one suspects that the battlelines are still drawn.

The reformers have a further strategy to escape the control of the bureaucratic careerists, and that is to push for greater democracy within the decision-making bodies of the institutes. According to Soviet administrative theory and practice an institute director has total power within the institute but he is obliged to present academic issues for discussion in the Academic Council, a collegial body of the institute's senior scientists (see Chapter 6). Some scientists have called for the Council to be given more extensive powers. Others are not satisfied with this, saying that since the Council is appointed by the director the extent of its powers is irrelevant. They wish to see the Council elected by the staff of the institute. Others wish to go even further, declaring that they want direct democracy with general meetings of the institute staff being given the powers that have been granted to the workforces of industrial enterprises in recent legislation. Again, as we saw in Chapter 3, limited concessions have been granted in the Academy, in that institute staff are now able to elect some members of the Academic Council.

We can conclude this discussion by asking whether any of the recent reforms represent an attack on the 'scientist manager' model. The essence of the model, and the problem with it, is that managers are given simultaneously great scientific and administrative power and are therefore subjected to the pressure of having to perform scientifically while essentially working as bureaucrats. The reformers want to attack this problem by decentralising both scientific and administrative powers to the research collective, but one suspects that they are not fully aware of the significance of the science–administration nexus. The evidence that researchers expect their managers to be good administrators, no matter what their scientific qualities, leads one to suspect that they will continue to want their managers to accept major administrative responsibilities, especially in solving the difficult problems of funding and supplies. But one even more strongly suspects that they would not be happy about having professional non-scientist administrators standing above them. This suggests a continuing commitment to the fundamentals of the present system.

Certainly the leadership of the scientific establishment has made it clear that it believes that the traditional approach must be maintained. This is strongly stated by Marchuk in the *Izvestiia* interview quoted above, but even more explicitly in an interview he gave in early 1987. There he stated that it was more important to have a good scientist than a good administrator running Academy institutes, since a poor administrator could always be backed up by administratively minded deputies.[67]

Conclusion

It appears that senior Soviet scientists are perhaps more concerned than their Western counterparts to keep publishing and to be seen to be contributing to the research process despite overwhelming administrative pressures. Whether this is overall a good or a bad thing is a matter of judgement. The pressure to keep publishing certainly can have negative consequences. Because they are scientists, and often scientists with well-established reputations, they wish to, and indeed are under pressure to, maintain their scientific reputations. This can lead to seriously negative efforts to maintain their positions through creating 'monopolies'. This is achieved by discouraging new scientific approaches, as well as such basically unethical behaviour as appropriation of research results and victimisation of potential competitors. A manager who no longer has any pretensions to a research reputation, such as a Western researcher who has chosen the 'administration stream', would be immune to such temptations. The ideal situation is obviously when a capable and honest scientist is also a good administrator. The Soviet system is well able to take advantage of such felicitous occurrences. But when fortune does not run its way, the system produces 'scientist managers' who at best impose an excessive conservatism on Soviet science and at worst create havoc through their determination to protect their own positions at all costs. There is clearly a high degree of awareness of these problems. But with the scientific hierarchy, and probably the scientific community as a whole, still committed to the traditional system, it is too early to say that a solution has been found.

Chapter nine

The staff

We have now looked at the organisations, institutions, and individuals that make science policy in the USSR. It remains to look at the people who actually do the research. I shall take as the starting point an examination of the job satisfaction of Soviet scientific personnel, and then shall look at the reasons for and consequences of the findings of that examination. Some might complain that this will ensure that the whole approach of the chapter will be a critical one, with the emphasis on the dissatisfaction of and problems faced by researchers. There would be some truth in such criticism. My only justification is that it appears to be precisely this dissatisfaction and the reasons for it which Soviet policy-makers are most concerned with at the moment. If there is going to be any significant change in Soviet science in the future it is most likely to be in reaction to these problems.

Job satisfaction

Unfortunately not much survey work is done, or published, in the Soviet Union which contains direct questions on job satisfaction. Most questions involving job satisfaction are in terms of satisfaction with particular and individual aspects of scientific work. However, a recent newspaper article provided an interesting summary of a survey of scientific–technical personnel which did apparently put the question in direct terms. It is reported that 90 per cent of those surveyed were basically happy with life. However, only about 50 per cent were satisfied with their job. The same percentage wanted to change their place of work, with over 15 per cent wanting to leave science altogether, 15 per cent wanting to move to teaching work, and 20 per cent wanting to move to another institute in the same field. Only one-sixth of those surveyed wanted their children to follow a scientific–technical career. It was stated that longitudinal studies showed that the desire to leave science is growing.[1]

That evidence, limited as it is, suggests a relatively high degree of job dissatisfaction. It is backed up by evidence of the falling prestige of science and in particular the declining attractiveness of science as a career among young people. For many years surveys showed overwhelmingly the high desirability of careers in science among children leaving school, their parents, and the population as a whole.[2] But in recent years concern has been expressed about a reduced interest in science among school leavers. The engineering teaching institutes have suffered in particular, with many of them, including some of the most prestigious, having trouble filling their quotas. It is said that applications for physics and mathematics faculties have also fallen sharply.[3]

There is perhaps sufficient evidence of relatively widespread job dissatisfaction in Soviet science to justify some consideration of possible reasons. There appear to be three main categories: dissatisfaction with what the Soviets call the 'socio-psychological atmosphere' in institutes; dissatisfaction with the more material aspects of work conditions; and a feeling that the rewards of a scientific career are inadequate.

The 'socio-psychological atmosphere'

In the survey mentioned above reporting a low degree of job satisfaction it was determined that over 33 per cent of those questioned were dissatisfied with the 'socio-psychological climate' in their institutes.[4] One of the obvious places to look for a source of such dissatisfaction is management style, although concern is also expressed about the problems of inter-staff relationships, an increasing anti-science atmosphere in Soviet society as a whole, and the unrealistic expectations and inadequate preparation of people moving into a research career.

Management

We devoted considerable attention to the problems of Soviet science management in the last chapter. We saw that because of a combination of the Soviet attachment to 'strong leaders' and having practising scientists in management positions there are problems of incompetent management and of managers resorting to unethical and corrupt procedures in order to maintain the appearance of successful research performance. We found evidence of plagiarism by research managers, plus considerable evidence of the victimisation of real or potential scientific rivals. It is hardly surprising that staff express dissatisfaction with these aspects of management. Thus in six surveys carried out between 1964 and 1974 in Novosibirsk, Leningrad, Alma-Ata, and Moscow, management problems were seen as contributing to conflict within institutes by 36.8, 50.8, 45.9, 41.8, 35.1, and 33.9 per cent of those discerning conflict as existing. The most common complaint within this category was about the bureaucratic methods of the manager, although 'imprecise determination of relative contributions to joint work' and 'lack of compatibility between planned projects and personal interests' also figured quite highly.[5]

The last factor reflects an apparent concern among many researchers that they have insufficient control over their own research work and careers. It is said that researchers who do not have their own laboratories have no easy access to funds or equipment for conducting their own research, and so find themselves little more than glorified research assistants or laboratory technicians (see Chapter 7).

It should be said that the data on these issues are not unanimous. There are surveys showing high degrees of satisfaction with management and the degree of research independence available, or alternative explanations for the phenomena observed. Thus a survey of two groups of researchers in a biology institute at Moscow University and the Institute of Cytology and Genetics of the Siberian Academy found that 90 per cent were happy with the qualities of their managers, and 90 per cent in Moscow and 95 per cent in Siberia were entirely satisfied with their relations with management.[6] This survey found that young researchers had little influence over the projects they were allocated (9.1 per cent chose their own project, 7.3 per cent chose their own but had it modified by their boss, 45.5 per cent had their project chosen by the boss although they had the opportunity to modify it, 29.0 per cent simply had their project set by the boss, and 9.1 per cent received their project by other unspecified means), and further that job satisfaction was higher among those who chose their own projects. Nevertheless, it was concluded that 'the data obtained permit us to conclude that the lack of independence in the choice of project is not a consequence of the authoritarianism of the manager, demanding the fulfilment of the institute's research plan. The main reason is the poor scientific preparation of the new researcher before his recruitment into the research collective.'[7] This is supported by a recent complaint about the lack of desire for independence displayed by young researchers.[8] We shall discuss the preparation and competence of researchers and the effect these might have on their job satisfaction later.

Inter-personal conflict

Surveys and memoirs show up problems of conflict among the staff of institutes. There is nothing unusual about that in any working collective anywhere in the world, and it is difficult to get any sense of how serious the problem is in Soviet science. Survey data that show losses of 6.04 per cent of working time through conflict in one branch institute suggest a less than catastrophic situation.[9] However, another survey of project managers found that they were certain of no conflict in only about 35 per cent of cases.[10] Just over half those in another survey considered it impossible to avoid conflict.[11] The Soviet press describes relatively regularly the havoc wrought by scandal-mongers with their endless complaints and often anonymous denunciations.[12] In one survey 40 per cent of researchers were unhappy with their colleagues professionally and personally.[13] The six surveys mentioned above allocated the blame for roughly half of all conflict situations to the character failings of researchers, with 'the unsociable character of the initiator of the conflict', 'researchers taking over the results of others' work', 'dishonest behaviour', 'laziness', 'hostility', and 'tactlessness' all scoring significantly.[14] Another survey refers to 13 per cent of conflict deriving from psychological incompatibility, but 33 per cent deriving from the rudeness, laziness, and dishonesty of individuals.[15] Another saw 87 per cent of conflict being based on the inability or lack of desire of researchers to conduct themselves in the appropriate way for collective endeavour, with 25 per cent of conflict situations arising from deliberate attempts to do fellow workers harm.[16] It should be mentioned here that another source of conflict which is treated more sympathetically in the press is the highly creative person with the difficult temperamental character. Such people regularly come into conflict with their superiors and colleagues, often it would seem because of their excessive devotion to their work, which one suspects usually means their intolerance of the laziness and lack of commitment of their colleagues. They are very often the targets of victimisation campaigns of both managers and rank-and-file staff.

Anti-intellectualism

Nobody likes being condemned as a 'lazy ivory-tower academic' who is responsible for all the ills of the world. If such attitudes are becoming widespread in Soviet society as a whole, one would expect them to have a negative effect on the prestige of science and the sense of satisfaction of scientists. In an earlier publication I suggested that anti-intellectualism was not a major phenomenon in Soviet society, although I did remind the reader of the anti-'bourgeois specialist' campaigns officially encouraged by the Bolsheviks in the early years of the regime.[17] While the official destruction of the bourgeoisie and the creation of a new socialist intelligentsia in the 1930s brought an end to these campaigns, the tension between communism as a working-class movement and one attractive to and appealing to all classes has never been entirely removed from Soviet ideology and practice. Under Brezhnev the leadership was said to be following a type of 'social contract' with the working class, at the expense of the intelligentsia. This was evident in pro-working class party recruitment policies, the narrowing of income differentials, and an ideological and propaganda stress on the ideological superiority of the working class.[18] All these forms of discrimination were keenly resented by the intelligentsia, and Gorbachev seems to be trying to remove them by opening up party recruitment policies and improving intelligentsia incomes.

However, he has also undertaken an anti-science campaign of his own, attacking corruption in the science sector and shifting the blame for the Soviet Union's poor technological performance from industry to the S&T sector. The first is part of a general campaign affecting the whole of society, but it has involved some very public disgrace of some previously untouchable science managers. The criticism of scientists for the failures of industry to keep up technologically is almost unprecedented.[19] Scientists are clearly concerned that these campaigns will add fuel to a growing anti-science atmosphere in the media and literature.[20] In a development not unknown in the West scientists are held responsible for the arms race, the destruction of the environment, etc.,[21] while Soviet literature, with a more documentary approach than is usual in the West, has dealt in some detail with the seamier sides of Soviet science.[22]

Unrealistic expectations

Any anti-science feeling that is evident in the Soviet media and literature must nevertheless be seen in the context of the traditional high status and prestige of science, as reflected in surveys and in literature. A number of commentators suggest that this reverential treatment of science in Soviet culture has created problems in that those growing up in that culture have gained an inaccurate and idealised picture of what scientific research entails. It is said that young people taking up scientific careers still have as their role model the traditional 'lone scientist', working independently on highly creative and stimulating projects. It is said that these impressions are often gained from biographical or fictional accounts of scientific research, which if they ever were accurate certainly are not today in the era of 'big science'.[23]

These unrealistic expectations are first seen in the very high value researchers or potential researchers put on creative fulfilment as their major motivation for undertaking a scientific career.[24] Ideologists regard such findings with mixed feelings, since they feel it would be more appropriate in a socialist society if a desire to contribute to the good of society rated higher than such 'individualistic' interests.[25] But others are more concerned that such unrealistic expectations lead to disillusionment and therefore a decline in morale and productivity once those suffering from them come up against the realities of modern day science, which are that young researchers in particular find themselves doing a large amount of routine and essentially boring work over which they have no control. Thus one survey showed that 33 per cent of young researchers had expected their work to be less boring when they took up their careers.[26]

We have already seen the dissatisfaction of researchers with the limited amount of freedom they have in choice of research projects and with management style and inter-personal relationships in the institutes. Such disappointments are particularly evident among young researchers, something which in the words of one group of Soviet sociologists contributes significantly to the difficulties they have in 'adapting' to research work.[27] However, one interesting fact that comes out of the work of this group is that there are very different responses on these issues between 'externally' and 'internally' motivated researchers. Internally motivated researchers are those who come into science with their own project in which they are already interested. They are presumably those who value creative fulfilment highly, whereas externally motivated researchers are those who in other sources are described as *sluchainye* (accidental), i.e. those who come into science because it promises security, 'an easy life', there was nothing else available, etc. It is those in the latter group who are less likely to find themselves in conflict situations and more likely to be satisfied with relationships within the collective. There is also a positive correlation between those who are in favour of closer supervision of their work and satisfaction with inter-personal relationships.[28] This adds reinforcement to the fears of the ideologists mentioned above that the individualist desire for creative self-fulfilment is inappropriate for work in a socialist collective. One must ask oneself, however, whether this greater adaptability of the externally motivated is in fact a function of a greater willingness to come to terms with low-performance institutions, on which more below.

Incompetence

The Soviet source on which we have been relying most here, while by no means ignoring the importance of 'socio-psychological' atmosphere in determining job satisfaction and adaptability, tends to see these problems as not deriving simply from poor management within the collective but from the lack of preparation of Soviet graduates for work in science. This is partly poor psychological preparation, in terms of the unrealistic expectations just described. But it is also a simple matter of inadequate training, so that people are being asked to do things which they simply cannot do, something which clearly has a negative effect on morale.

The achievements and individual reputations of many Soviet researchers demonstrate conclusively that competence is something which is by no means excluded in Soviet conditions. However, a number of questions have been raised by both Soviet and Western commentators about the overall level of competence of Soviet researchers.

The first concerns their education and training. It is not my purpose here to analyse deeply or comprehensively the Soviet education system, but some details are required. The Soviet higher educational system can be roughly divided into three categories of institution. There are the universities, offering a full range of courses including the fundamental and social sciences; the polytechnical institutes, teaching a range of technical and applied science disciplines; and the branch teaching institutes, offering both theoretical and applied instruction in specific disciplines.[29] In 1986–7 there were 896 *vuzy* in the USSR, of which 65 were universities, 64 were polytechnical institutes, and over 600 were branch institutes.[30]

Undergraduates in all *vuzy* work towards their diploma (*diplom*), the equivalent of a Western Bachelor's degree. This requires the taking of a range of courses, usually over four or five years, with regular oral examinations marked in the usual Soviet way out of five. In the final year the student is required to complete a *diplomnaia rabota*, a substantial essay based on research and roughly equivalent to the honours sub-thesis in an Australian honours year. An excellent student interested in a research career can then proceed to prepare for the candidate of science degree by enrolling in a postgraduate programme, the *aspirantura*. Some preliminary preparation as a *stazhër* (perhaps best if not entirely accurately translated as an intern) for a year or two in a research institute is usually required. A full-time *aspirantura* can then be undertaken, either in a research institute or a teaching institution. This involves some course work and a demonstration of an adequate knowledge of Marxism–Leninism and a foreign language, as well as a research dissertation. When completed the student is granted a candidate's degree in the relevant discipline. The degree can also be gained part-time, and is roughly equivalent to a Western Ph.D. The final degree available to Soviet scientists is the doctorate. In recent times this has not involved any formal programme of work but has been awarded on the basis of work published in the course of one's research career. In practice it appears that usually a scientist writes a monograph with the specific intention of submitting it for the doctorate. The doctorate is usually granted to senior well-established researchers. There has been mention in recent legislation of establishing a *doktorantura*, presumably meaning a formal programme equivalent to the *aspirantura*.

There is some difference of opinion over the system's adequacy, with some Western commentators seeing it as superior to that offered in the West but others, both in the West and in the Soviet Union, considering that its excessive emphasis on rote learning leaves analytical skills undeveloped.[31] The education provided is also seen as overspecialised, particularly in the branch *vuzy*. These have come to provide an increasing percentage of research workers, at the expense of the more general and theoretical educations provided by the universities.[32] There is also concern at the slowness with which educational systems and programmes are restructured at times of rapid scientific–technical progress.[33]

But the main concern appears to be not with the actual material taught in the education system but with the lack of practical preparation for research work. The concentration of Soviet research in independent institutes means that *vuz* students have little direct contact with active researchers and the equipment they use. In 1968 a survey of department and laboratory heads of the Pol'zunov Central Steam Turbine Institute in Leningrad revealed their opinion that less than half their subordinates were adequately prepared for research work.[34] A recent survey of young researchers found that 23.6 per cent had no knowledge of the problem they were working on before they came to the institute, 30.9 per cent had little knowledge, 38.2 per cent knew of the problem only in general terms, and only 7.3 per cent knew the problem well. Just over 29 per cent claimed to have practically no knowledge of the necessary research techniques, 34.5 per cent had a patchy knowledge, 30.9 per cent had a good knowledge with only minor gaps, and 5.5 per cent considered their knowledge complete.[35]

The problem is not one that has ever been completely ignored. It has always been common for top-level researchers simultaneously to hold teaching posts in *vuzy* (*sovmestitel'stvo*), and it is usually openly stated that the main motive has been to allow researchers to pick up talented youngsters early in their education.[36] Some institutes even go talent-spotting in schools.[37] Many research institutes take all their new researchers from a limited range of *vuzy*,[38] and some have even set up departments (*bazovye kafedry*) in those *vuzy*.[39] These close links between *vuz* and research institute make it possible for educational programmes to be tied closely to the needs of the institute. It also makes possible the operation of the *stazhirovka* system, in which recent graduates are allocated training posts in a research institute before being allowed to go on to postgraduate work. Often undergraduates in the final year of their degrees are allocated supervisors for their *diplomnaia rabota* from research institutes.

Despite all this it is considered that the research preparation of *vuz* graduates is still inadequate, and that links between *vuzy* and research institutes will have to be further strengthened.[40]

It seems possible that such increased links will accentuate an existing problem of overspecialisation within the research sector itself. Indeed it could be argued that the overspecialisation within the education system is a result and reflection of the overspecialisation in the research sector.

The traditional non-project approach to research planning and structures, the 'sectional' or 'departmental' jealousies which are evident in the branch sector in particular, and the low level of staff mobility from institute to institute have produced research institutes that are highly discipline specific and specialised. Some Soviet analysts have presented data which they consider demonstrate inadequate cooperation between institutes on individual projects, and even between the subunits of individual institutes.[41] This overspecialisation is seen as leading to an inability and unwillingness to take on the inter-disciplinary projects that are so important in modern science and technology.

Concern is also expressed about the low numbers of researchers with advanced degrees, particularly in the branch sector. Data from the late 1960s showed that over half the senior research personnel in branch institutes had no advanced degree, with there being just one doctor of science per institute on average. The situation was said to be particularly bad outside the metropolitan cities.[42] A 1986 article on the problem mentioned two institutes of about 1,600 staff each with a single candidate of science.[43] Table 9.1 presents data comparing the situation in the Academy and branch sectors.

Table 9.1 Advanced degrees in the Academy and branch sectors (000s)

	No. of research workers	*Doctors of science*		*Candidates of science*	
		No.	*%*	*No.*	*%*
Academy					
1970	85.1	5.7	6.7	34.0	40.0
1980	125.6	9.3	7.4	55.6	44.3
1985	137.3	11.4	8.3	64.9	47.3
1986	141.9	11.8	8.3	66.2	46.7
Branch					
1970	420.0	5.1	1.2	62.0	14.8
1980	608.3	8.5	1.4	109.4	18.0
1985	662.2	9.6	1.4	125.7	19.0
1986	661.1	9.7	1.5	127.7	19.3

Source: *Narodnoe khoziaistvo za 70 let*, p.62

Some of the problems of Soviet higher education have been the focus of the series of decrees that appeared in March 1987 (see Chapter 5). They contained the usual statements on the need for greater on-the-job training, including a call for setting up 'study–scientific–production complexes' in major enterprises where students could get experience of industrial innovation. Somewhat more interestingly, provision was made for special teaching and training programmes within the *vuzy*, including the possibility of adding an extra year to the degree programme for specially selected students destined to go on to research careers. Other points were the setting aside of 15 per cent of teaching time specifically for the 'operational expression' of the latest achievements in science and technology, a reduction in the amount of lecture time to give students more time for individual study, a reduction in staff:student ratios, and getting industrial ministries to pay for training of specialists in their fields. More stress is to be put on full-time *aspirantura*, and a formal research programme leading to a doctor's degree, the *doktorantura*, is also to be established, with there even being provision for this programme to be done full-time. Finally, concern was expressed about past barriers put in the way of full-time researchers holding simultaneous teaching appointments (*sovmestitel'stvo*). The encouragement of such dual appointments is seen as necessary to improve the quality of teaching.

The decrees are clearly a response to perceived problems. They are not radical, but if implemented could bring about useful changes. Many are being put forward for by no means the first time. It remains to be seen whether the Gorbachev administration will approach implementation with more determination than has been evident in the past.

Poor conditions

The second source of job dissatisfaction in Soviet institutes is a perception on the part of researchers that they are working in poor conditions. Soviet science suffers a general lack of infrastructural facilities, whether in working accommodation, equipment and day-to-day supplies, the developmental facilities required to make good use of research discoveries, or social amenities for research staff.

Social amenities

As far as social amenities are concerned, the Soviet movement in recent years towards the 'science city' concept, i.e. concentrating scientific institutes in single areas, often on a 'green field' basis, has accentuated the problems that arise when the basic work collective is responsible for providing many social and basic living amenities for its staff. The problem is that the science sector does not have the funds, and does not give priority to social infrastructure, meaning that institutes are unable to provide flats, in particular, but also holiday, sports, and other facilities. Thus we see complaints of the critical shortage of housing for staff of the institutes of the Academy's Department of General Physics, including 500 on the waiting list at the Leningrad Institute of Nuclear Physics which is situated in 'science city' type circumstances in Gatchina, outside Leningrad. There is a 12–15 year wait for flats in the 'science city' at Chernogolovka outside Moscow.[44] These difficulties make it hard to recruit staff and worsen morale. They can lead to labour immobility since once on the waiting list or in their flat people are most reluctant to start all over again. The president of the Academy Marchuk recognises that the housing problem is something which endangers his plans to increase radically the turnover of personnel and the intake of young researchers into the Academy as a whole.[45] But it can also lead to excessively high mobility. The Far Eastern Science Centre of the Academy of Sciences, for example, over the last 16 years has had a net outflow of candidates of science, while 35 per cent of young researchers sent to the Centre leave each year. Isolation is clearly a factor, but considerable blame is also put on poor living conditions and social amenities.[46]

Working conditions

There are also regular complaints that institutes themselves have far from adequate premises. For example, the new Institute of Problems of Information Science finds itself spread among 17 different buildings in Moscow.[47] A problem that such institutes face is the extreme difficulty involved in getting planning permission for new buildings in the metropolitan cities, even assuming that it is possible to get funding allocations and the projects included in the plans of construction organisations. In 1983 academician B.S. Sokolov complained that in the last 50 years only one building in Moscow had been purpose-built for scientific research in the branch network.[48] Construction delays are a notorious problem in the Soviet Union, and science projects seem to suffer at least as much as those in other sectors of the economy. The Academy of Sciences has its own construction organisations,[49] but this does not seem to help.[50] A recent Politburo meeting included the science sector among those that should get priority access to new building in Moscow.[51] It remains to be seen whether that will ease the situation in that city at least.

Within the institutes themselves one finds ridiculous situations such as senior research workers not having a desk to themselves, to say nothing of a separate office, even in prestigious Academy institutes.[52] The Academy continually complains of its lack of photocopiers, or shortages of paper, ink, and spare parts that render inoperative those machines that they do have. The Academy's 'Physics and Mathematics Literature' publishing house did not possess a photocopier in 1986.[53] Access to foreign journals is another serious problem.[54]

A separate problem is that of the supply of scientific equipment, apparatus, and instrumentation once the basic buildings and working facilities have been provided. It is generally considered that Soviet scientific equipment tends to be in short supply, is of uncertain quality, and can only be obtained by going through cumbersome bureaucratic procedures. Investigations carried out in the mid-1970s showed that only 25–50 per cent of the planned demand for equipment was satisfied, while planned demand covered only 40–50 per cent of actual demand. Figures for the early 1980s for the Belorussian Academy of Sciences were similar.[55] The supply of apparatus, isotopes, chemical compounds, etc. is planned, which means that orders have to be submitted as much as two years in advance. In one institute it was reported that between one-third and one-half of working time was spent travelling around the country trying to obtain essential supplies. Another survey gave a figure of 15 per cent of working time spent on obtaining supplies.[56] Problems with equipment are rated as one of the main sources of dissatisfaction among researchers, while failure to sort out supply problems are a major source of staff dissatisfaction with managers (see Chapter 8). The Harvard survey found a number of respondents prepared to make the extraordinary claim that the ability to manufacture alcohol was one of the most important factors for a successful research establishment, since it is a universally acceptable unit of exchange even for scientific equipment.[57] That might explain why of 1,000 litres of alcohol delivered to the Academy's Industrial Laser Research Centre in 1986 500 litres went missing.[58]

Ancillary staff

Shortages of ancillary staff are strongly felt, and are seen as contributing to the amount of time scientists are forced to divert from research. Statistics on research:ancillary staff ratios are far from consistent, but generally show a rapidly worsening situation over recent decades. One set of global statistics shows the ratio declining from approximately 1:4 in 1960 to 1:3 in 1965 and 1:2.5 in 1970.[59] These data roughly parallel those for the Ukrainian republic of 1:2.53 in 1965, 1:2.45 in 1970, and 1:2.20 in 1975.[60] A recent figure, derived from a survey, which is specifically the ratio of scientific workers to laboratory assistants (*laboranty*) gives the startling figure of 7:1.[61]

This source stresses that the situation has become much worse in recent years as ancillary staff have been the first to suffer under pressure to reduce staffing levels.[62] The very poor pay rates for scientific ancillary staff compared with those to be obtained in industry add to the difficulty.[63] The result of the worsening situation is that researchers are forced to devote considerable time (15–30 per cent of working time according to one source) to menial and manual tasks.[64] One commentator specifically links the situation to the decline in the prestige of research careers.[65]

Administrative duties

There are still other activities which divert scientists from their research, and lead to a sense of frustration and even demoralisation. The first is administrative work. Soviet sources complain about the high levels of working time lost through, among other things, administrative duties. A Ukrainian Academy of Science survey found that junior research workers spent on average one hour per day on administration, senior research workers two hours, and department heads three and a half hours.[66] A lot of this time is spent on planning work (leading researchers spend up to one- third of their time on planning, according to one source),[67] a lot on the preparation of documentation and having it circulated and signed,[68] in meetings,[69] and on simply waiting to see one's superior.[70]

These losses of time on administrative duties seem comparable with Western experience, although rank-and-file researchers perhaps rely to a greater degree on their superiors to handle administration for them. This is to a considerable extent the result of the lack of individual research grants in the USSR. It is also, of course, a function of the relative lack of rank-and-file participation in management decision-making in Soviet research institutes. It will be interesting to see whether any democratisation that might take place will change this.

'Community service'

The final area of difficulty is *obshchestvennaia rabota*. This term, directly translated as 'social work', covers a whole range of tasks that might be better described as 'community service'. It includes membership of various elected bodies, both within the institute (trade union committees, party commissions, etc.) and outside the institute (the local soviet, local party organs, etc.). It also includes service in such bodies as the *druzhina*, a sort of civilian militia which assists the uniformed police force, as well as lecturing at agitation and propaganda meetings, often organised by the *Znanie* (Knowledge) Society. These sorts of duties tend to affect senior researchers and managers more than rank-and-file staff, although party members at all levels are likely to find themselves particularly liable to such service.[71]

The most notorious form of community service is going '*na kartoshku*' ('on the potatoes'). This is a generic term for the practice of local officials forcing local employers to provide labour for a whole range of menial and manual tasks. These tasks are usually agricultural, particularly at harvest time, but can also include working on construction sites or in warehouses. In the past such duties were sacrosanct, with failure to perform them being grounds for dismissal.[72] In very recent times, however, the practice has been subject to public criticism, and some general data on time lost have been published. It is said that in 1987 research institutes lost 8.7 million worker-days (4.2 days per worker) and design bureaus 5.0 million (4.7 days per worker).[73] In that year every able-bodied member of the Scientific-Research Computing Centre of the Academy's Pushchino Biology Centre lost on average 20 working days and the work of the institute had been totally disorganised for at least three months.[74] Agricultural assistance is obviously a very sore point, but given the manual nature of agricultural work in the USSR and the shortages of permanent labour, it is one which will not be easily cured.

We can conclude this section by presenting some data on overall losses of research time to all forms of non-research activity. Different surveys show various amounts of time spent directly on research work, from as little as 16 per cent of the working day in one Leningrad survey to about 60 per cent in another survey of 700 researchers.[75] All surveys show that the higher the post occupied the greater the amount of time lost. The time spent on non-research activities in a survey of 29 Leningrad research and design institutes in 1967 was as follows: department head, 29 per cent; deputy department head, 32 per cent; senior research worker, 18 per cent; junior research worker, 12 per cent.[76]

Material rewards

The final area of dissatisfaction among research staff is material rewards. A significant number of scientists say that they took on a scientific career primarily for material reasons,[77] but one would expect that even those less materially oriented would also be concerned at poor financial returns. The main areas of concern are wages and promotion prospects. In both these areas there have been problems over the last two decades. Throughout the 1960s and 1970s scientists slipped from the highest paid to the fourth highest paid occupational category and their pay advantage over the national average was reduced virtually to zero. It is true that in 1976 salaries were increased and scientists were given greater opportunities to earn bonuses. Nevertheless their overall position in the pay league did not improve, and the feeling was strong that they were discriminated against compared with blue-collar and even unskilled workers.[78] The comparative data on the average monthly wages of industrial personnel and staff in the science and scientific services sector (in rubles) are as follows:[79]

	1965	1970	1975	1980
Industry	104.2	133.3	162.2	180.5
Science	120.6	139.5	157.5	179.5

At the same time that pay levels were suffering, promotion prospects were also taking a serious turn for the worse. The enormous expansion in the research network that had been evident through the 1950s and 1960s slowed very considerably in the early 1970s. There was little movement of established researchers out of science but at the same time the number of new management positions was greatly reduced. This led to a remarkable ageing of the Soviet scientific community and great frustration for those who found themselves at the bottom of the pile.[80] One survey showed that job dissatisfaction is greatest among the middle-aged (35–44 years), presumably when the choked promotion ladder becomes most frustrating.[81]

Productivity in institutes

This has been a formidable list of sources of dissatisfaction among Soviet researchers. It is difficult to determine how widespread such problems are and how keenly they are felt. However, even if they exist only to a relatively minor degree one would expect to see them having some effect on the quality of Soviet research. We have already seen the apparent decline in the status and desirability of a research career in recent years. Here I want to look at the effect on how research is done. The most obvious result is a very low level of productivity as reflected in the behaviour described in my favourite quotation from the Soviet press. It comes from a letter written anonymously by a rank-and-file researcher in 1973:

> You will not believe this, but our working day proceeds thus: the men talk about football or the cinema or play ping-pong. Sometimes they do some work. The women go shopping, drink coffee, knit, sometimes write something. While I write this letter there are three people in the room beside me. One young man, a candidate of science, is sitting at his desk looking through foreign magazines. He copies out various items on cars, aeroplanes, sport and so on – it's his hobby. He copies it out or clips it and files it away. An 'encyclopaedist'. Two others – not candidates – are writing something. Two are standing on the stairs smoking. One woman, a candidate, has gone off on trade union business. I don't know where the rest are. They wander in and out. . . . They work very little. I would guess about two hours a day. . . . I could change jobs. But in my field there are only two other institutions, and I've spent some time in both. I didn't like them. Everybody fought between themselves and they all tried to palm their work off onto others. Here we have such a friendly collective. They're very nice people. Sometimes our room is full of laughter.[82]

This picture strongly suggests organisations of low morale and poor management. Western studies have suggested that the consequences of such conditions are far more serious than simply reduced productivity. Positive harm is done by, at best, lack of due care in the preparation, conduct, and recording of experiments, and at worst deliberate falsification and sabotage. This is particularly likely to be a problem where the person doing the work feels little responsibility for the results of his activities or feels that he will get little credit for any good results he might achieve.[83] Those are the conditions which seem to rule in a significant number of Soviet research institutes, and we can only assume that the quantity, quality, and reliability of Soviet research suffers as a result.

Attempted solutions

The current political leadership is clearly very dissatisfied with the return they are getting on their investment in the science sector. They are prepared to attack the problem on a macro-scale by increasing funding and investment. On the micro-scale their approach has had that same element of the carrot, but the stick has also been much in evidence. Everything done at this level has been directed, as has the whole *perestroika* programme throughout Soviet society, at raising the morale of the population following its collapse in the final years of the Brezhnev regime. Even the stick is designed to improve morale by impressing on the majority of Soviet workers who are honest and conscientious that they will no longer have to carry those who are not.

There have been three main aspects of the Gorbachev strategy. Firstly there has been a purge of personnel at all levels of Soviet science, a purge directed against corruption, incompetence, and simple old age. The most spectacular and seemingly thorough-going purge of corrupt scientists has been in the Kazakh Academy of Sciences. The Kazakh Academy was the subject of a full investigation by the Committee for Party Control, the party's chief investigatory and disciplinary body. The president A. Kunaev (perhaps not coincidentally the half-brother of the also deposed first secretary of the Kazakh Communist Party) was replaced for his dictatorialism, cronyism, drunkenness, and generally lax and corrupt management.[84]

As we have already seen, the president of the central Academy, A.P. Aleksandrov, has also recently been replaced, although in this case the motivation seems to have been simply age and resistance to change rather than corruption. The introduction of a compulsory retirement rule in the Academy and *vuzy* has led, as has also already been described, to the removal of a considerable number of institute directors and other senior science managers. Although we have no precise data on turnover at the top of the branch R&D sector as a whole, one strongly suspects that it has been considerable there also.

The purge has been carried out at lower levels too. Since about the mid-1970s the *pereattestatsiia* (attestation) system – what might be best translated as 'efficiency review' – has been quite openly used to reduce staff. Each researcher has his work examined by a commission, usually made up of his superior and trade union and party representatives, which has the right to recommend promotion, demotion, or even dismissal (with demotion or the threat of dismissal usually leading to resignation). In the past the system was often criticised for the weakness of its effect, mainly the result of the reluctance of managers to get tough with their subordinates. But since Gorbachev came to power the whole science sector has been subject to a new round of reviews. The staff reductions attained as a result have been significant without being spectacular. A preliminary application of the system in Leningrad led to a 6 per cent reduction in staff, while a review of Academy personnel led to 2 per cent of staff being dismissed or resigning and 3.2 per cent being demoted.[85] Based on hearsay evidence, a strong sense of nervousness has spread through research institutes.

The stick aspect of efficiency review and purge is obvious. However, they are also designed to contribute to the carrot in two ways. Firstly they are designed to reassure honest and conscientious researchers that the ballast will now be removed; secondly they are designed to contribute to the earnings of those who remain, by opening up greater opportunities for promotion and by allowing savings on wages through staff cuts or demotions to be distributed among remaining staff.[86]

Some published cases suggest that there is scope for increased pay and bonuses in conditions of widespread dismissals, demotions, and retirements,[87] although a continuing effort to reduce the proportion of the science budget going to wages means that any increases will have to be found from internal funds. One must also remember that one of the major sources of wage savings is through the rationalisation of institute structures. While in the short term this might provide the funds for wage increases, in the long term it can only further reduce promotion possibilities.

There are also problems with the effort to increase morale through efficiency review and purge. Firstly, as far as the psychological effect on staff is concerned, this relies on the perception of staff that the system is being applied fairly. Hearsay and published remarks suggest that efficiency review has in fact been a boon for the established scientific authorities, the very people who were supposed to be its victims. They have used the system, in the same way as they always have, to get rid of their opponents. A recent article in *Pravda*, after referring to the reversal of the unfair dismissals of a number of researchers, made the statement that 'it happens that the initiative is seized by those for whom efficiency review is a personal weapon, a chance to solve "surgically" old scientific and non-scientific disagreements, to remove an opponent or competitor, and most often of all to rid oneself of a critic, that "difficult" character who for so long has caused irritation'.[88] If it is true that, as some have suggested, this problem has become worse under *perestroika*, the efficiency review system will in fact have a negative effect on morale.

With regard to increased remuneration, the question is always the same – will bonuses be paid equally to everyone regardless of performance, or will they go only to those who are truly productive and efficient? Despite regular official campaigns against *uravnilovka* – the equalisation of wages and bonuses – the pressure within Soviet work collectives to allocate increases according to that principle is always enormous. Indeed one suspects that the effects on morale of serious efforts to apply a truly differentiated approach would make most research managers reluctant to contemplate it.[89]

The second approach to introducing discipline at institute level has been to put institutes on *khozraschet* and self-financing. The details of this change, such as are available, have already been presented in Chapter 7. The basic idea is that institutes have to pay their own way, through planned tasks with central funding but also to an increasing degree through contract research. If they cannot cover their costs with such receipts bank credit will be made available initially but eventually they will have no choice but to go into liquidation. Again the stick is pretty obvious, a stick which is applied to both management and staff – both will have to work harder and more efficiently if they are not to find themselves without a job. But there is again a carrot included, in that institutes will have greater freedom to retain and distribute profits among staff (in the form of both increased bonuses and improved working and social amenities). It is still too early to say how effective either the stick or the carrot will be.

The final strategy has been democratisation. Work-place democracy has been clearly championed by Gorbachev as a way of putting pressure on established management personnel and practices. Thus it is a stick as far as management is concerned, but a carrot for rank-and-file staff. As we have seen in the discussion of democratisation in the Academy in Chapters 3 and 8 the signs are strong that democratisation in the science sector is being used with these goals in mind. However, it is too early to say whether the democratisers will be allowed to gain the upper hand, over the short, medium, or long term, or whether institutes will be any happier places even if they do.

One suspects that in the final analysis the established management figures will gain the upper hand, and that even if they do not those that replace them will find themselves operating in a similar environment and facing the same types of pressures. One suspects, admittedly on the basis of very little evidence, that Soviet managers will continue to be mediocre and authoritarian. At the lower levels, as anyone with direct experience of Soviet working collectives knows, low-productivity work habits are deeply engrained and indeed are barely perceived as being low productivity. It is not altogether surprising that there is evidence of strong and effective resistance to or scepticism about *perestroika* among the scientific intelligentsia, although the intelligentsia is considered to be one of its major beneficiaries. This is particularly evident in the Academy, long considered as one of the more dynamic Soviet institutions but at present displaying determined opposition to change.[90]

One wonders, paradoxically given the general Western picture of the Soviet Union, whether the authorities are willing or able to be tough enough with the rank and file. A popular concept in the study of regime–population and management–staff relations has been the 'social contract'. This can be summed up most crudely in the phrase 'we pretend to work and they pretend to pay us'. It contains the idea that the general working population does not receive or expect overly generous treatment from their superiors in terms of living standards or financial remuneration. In return they expect to be given security of tenure despite low-productivity performance. One would not want to lay too much stress on the 'social contract', but to the extent that it exists at all it produces a situation in which an agreement is reached between management and the vast majority of staff not to rock the boat. Those who try to stir things up, particularly by working too hard, are a small minority equally disliked by staff and management.[91] They are the victims of the management authoritarianism we saw in the last chapter. One doubts that Gorbachev's measures have been enough to break the 'social contract'. If that is so purges and discipline might reduce the amount of ping-pong played in Soviet research institutes but are perhaps unlikely to increase significantly the amount of good research being done.

Chapter ten

Main themes and current developments

As I said at the very beginning of this book, I do not claim the expertise in Western science to have made this study explicitly comparative. My guess is that at first glance the differences between Soviet and Western science might appear major: there is the relatively small amount of 'academic' research done in universities, meaning that the bulk of basic research is done by full-time researchers without teaching responsibilities; the total control by the state of industrial research, rather than by private corporations; a highly centralised planning system; the determination of research managers to retain a research reputation rather than accept their fate as administrators; and a highly authoritarian management style. (This is to say nothing about the greater political control that is usually considered to be imposed on Soviet science, something that has not featured importantly in this book although not because I consider it insignificant.)

On further reflection some Western scientists might conclude that the differences are not after all so great: that many top researchers consider it important to maintain links with teaching institutions; there is in practice little difference between the state bureaucracy of the USSR and the bureaucracies of the largest Western corporations, which are so often tied into the state system through military and other contracts anyway; the central planning system is in practice infinitely flexible, and frees many researchers from the horrors of the research grant system; the Soviet 'scientist manager' concept is a bit of a myth and increasingly recognised as so; and the reports of authoritarian misdeeds in Soviet science are probably not characteristic of Soviet science as a whole and have their parallels in the West anyway.

My conclusion, however, if I were forced to come to one, would be that the differences are significant, and that they have led to poorer performance by the Soviets. This would seem also to be the view of many Soviet leaders and specialists. They are increasingly scathing of the performance of their scientific sector, and many of their proposed changes could be seen as looking towards Western models, albeit vaguely. These developments have increased in intensity in the Gorbachev era. This final chapter will combine a concluding summary of the features and problems of Soviet science and an analysis of the most recent changes introduced under Gorbachev. The risk of instant obsolescence in the latter enterprise is very high.

The examination in the earlier part of the book of the structures of Soviet science policy-making and management suggested that Soviet scientists have a considerable degree of influence over these processes. Although scientists do not have top-level political positions or power there are major bureaucratic structures – GKNT, the Academy of Sciences, the *vuz* management organs of the State Committee for Education, the S&T organs of the ministries, etc. – which allow scientists to participate in formal consultative and coordinating procedures. Further, high-status scientists, including often those that do not hold the top management positions, have informal access to top political leaders. The inevitable *de facto* decentralisation of the science planning and management system further increases scientists' influence over the details of planning and management.

These opportunities for access and influence arise in the context of traditionally a high degree of ideological and popular prestige afforded the science sector. Science is recognised throughout the society as an important contributor to economic and social development.

Prestige plus policy influence have meant that the science sector has been able to establish itself as a secure and relatively well-funded part of the Soviet socio-economic system. It has responded to the favours granted it by performing quite impressively. Soviet science occupies a legitimate place on the world stage, the nation has industrialised to a large extent on the basis of its own R&D, and on occasions it comes up with spectacular scientific and technological coups.

Despite this there is strong evidence that Soviet science and technology lags, sometimes seriously, behind the West. Its fundamental research is isolated from the international community in a way which suggests that a problem of quality is involved. The backwardness of Soviet technology is if anything more obvious, in terms of the quality of Soviet industrial output. Soviet commentators have been increasingly prepared to say that this backwardness is the result of poor research and not of problems at the production end.

Indeed in recent years there have been scathing attacks, from Gorbachev down, on Soviet S&T performance in general and on specific programmes and individuals. What has been particularly worrying for Soviet scientists is that these official attacks come against the background of an apparently significant popular rejection of science. Science, in a development not unknown in the West, is blamed for environmental degradation, a denigration of spiritual values, and the rise of militarism. Science is losing its popularity as a career, for these cultural plus material reasons.

The official attack on science is to some extent part of Gorbachev's desire to shake up all sections of society. But there are also specific criticisms which suggest that the science sector is seen as a major target in its own right.

These criticisms include attacks on the enormous waste of funds in major programmes which have been undertaken without any serious outside examination or control. There is then a more general concern with the lack of financial and material responsibility in the science sector as a whole. This is seen as deriving from the traditional central planning system, in which operational units are given *de facto* control over planning and spending but without any incentive to maximise performance. Science has also been isolated by the planning system from production. As a result the science sector has become a closed world operating according to its own self-defined parameters of performance.

The real problem arises from the fact that the lack of material responsibility is combined with a lack of moral responsibility. One of the most disheartening features of Soviet life in the final years of the Brezhnev era was the almost complete collapse of morale and morality at all levels of society. In the science sector the decline was hastened by a scientific élite which to a significant extent was corrupt and authoritarian. Some were corrupt in the purely material sense, but most were corrupt in a more intellectual sense. Their main interest was in maintaining their scientific 'monopolies', either because psychologically they were unable to give them up or because they recognised that their administrative power rested on them. In fundamental research these monopolies might represent out-dated scientific theories (Belousov's 'fixist' geology); in industrial R&D they were often the conservative technologies which were traditional in the branch and best suited the interests of the dominant production managers.

Often this state of affairs suited the rank-and-file researchers of the institutes. The 'wheeler-dealer' type managers were often better able to look after the interests of the collective, through the deft use of bureaucratic skills, than intellectually more capable and honest 'ivory-tower academics'. But more importantly, most researchers found it relatively easy to do a deal with such managers, particularly once the initial excessively idealised view of the joys of scientific creativity had been exposed to the hard light of day. The deal was that researchers would accept and help maintain the monopoly position of the manager, but in return they expected a secure and relatively relaxed existence. This is the scientific version of the notorious 'social contract' ('we pretend to work and they pretend to pay us') which many consider to have ruled in the Brezhnev era in particular. Those who refuse to accept the contract, i.e. those who attack the monopoly position of the manager or the low-productivity work habits of their colleagues, find themselves subject to the full rigours of Soviet scientific authoritarianism.

Gorbachev has demonstrated very conclusively that he is prepared not just to talk bluntly about the problems facing the Soviet Union but also to do something about them. The science sector has not escaped his attention. He has not been purely negative in his approach, since he has offered substantial improvements in funding. He has also moved strongly in the direction of a decentralisation of the planning system, essentially through the introduction of 'full *khozraschet* and self-financing' and treatment of scientific output as a 'commodity'. One intention is to reduce the information overload which creates serious distortions in the planning process. But more emphatically the intention is to force operational units to accept the financial and material responsibility which should accompany their considerable control over planning, funding, and management. A carrot incentive is involved, in the form of the opportunity to make more money; but there is also a very strong stick incentive, that if you cannot pay your way you go out of business.

Gorbachev has tried to match the drive for material responsibility with one for moral responsibility as well, through an attack on the corruption of management and the management–staff deal. He has purged not only the blatantly financially corrupt, but also the intellectually corrupt and bankrupt. The turnover of top-level science managers, after a slow start, has become spectacular. At the same time there has been a strong push for the democratisation of science management procedures, primarily through the election of science managers. It is hoped that this will give the honest and conscientious members of the rank and file the opportunity to exert their influence over the scientific process.

The changes have been spectacular and traumatic for many. However, given my generally negative evaluation of Soviet science management, it is perhaps not surprising that I remain sceptical as to the likely success of the reforms. There is the question of whether the Soviet Union will ever be able to overcome its relative lack of resources, given that the West, if it can be seen in any way as a single entity, will be able for the foreseeable future to outspend the Soviets comfortably. At the moment the science budget looks secure, even at a time when major macroeconomic budgetary difficulties are admitted in ever more worried terms. But there does not seem to be room for further major increases.

But perhaps more important than funding are the issues of material and moral responsibility – will the allocated resources be put to the best possible use? When we look at the efforts to increase material responsibility, through the decentralisation of the planning system, we are not surprisingly struck by the degree of bureaucratic inertia and downright opposition. As in previous attempts at decentralisation the opposition comes not only from senior bureaucrats who stand to lose their powers but also from operational managers who stand to gain only increased responsibilities. One can detect signs of near panic among institute managers as they go over to *khozraschet*, and one wonders whether they have the nerve, knowledge, and competence to cope. Certainly the natural constituency for decentralisation would seem to be a small one.

There is the further problem of what to do with those units and individuals who are not able to pay their way. The issues of unemployment and bankruptcy are major ones in the Soviet Union today. They are discussed in the press, but with a nervousness which suggests that no one really wants to bite the bullet. One suspects that the science sector has enough slack that a truly tough approach to *khozraschet* would have devastating consequences. One wonders whether the leadership has the will.

The final question is how *khozraschet* will work if the interest of users in the output of research institutes does not increase. Of course in the case of fundamental research there might not be a user. Here scientists are being assured that the funding of fundamental research is safe, even if there will be more competition involved in deciding who gets it. At the moment one can perhaps accept those assurances at face value. The problem for applied research is more difficult. Applied research does have users, but traditionally they have been notoriously unwilling to play their role, beholden as they are to production plan fulfilment at all costs. Many fear that the stricter application of *khozraschet* in industry will only increase that unwillingness. The official line is that with increased competition in industry enterprises will be forced by market pressures to innovate, but many commentators are far from reassured. There does seem to be a very real danger that even good R&D will be unable to find ready customers.

With all these question marks over decentralisation it is not surprising that various escape routes have been left open. The most obvious is the *goszakaz*. It is certainly far too early to say that centralising pressures will not reappear in the future.

It is harder to judge the likely success of attempts to increase moral responsibility, since that requires an assessment of the psychological state of mind of the scientific community. One does, however, get the sense that the measures taken against corrupt, conservative, and unproductive management and staff underestimate the strength of the management–staff deal. The purges of top managers have been impressive, although it is too early to say that the replacements are a major improvement. The purge of unproductive rank-and-file researchers, through 'efficiency review', has clearly been far less successful, with the review indeed being used to a large extent to strengthen the authoritarian positions of managers. Democratisation is as yet in its infancy, but it will work in the way desired only if there are a lot of people in the institutes who want change. Some scepticism on that account is in order.

My rather pessimistic evaluation of Gorbachev's reforms is essentially in terms that they are up against too powerful an opponent rather than that they are in themselves wrong. The evaluation is therefore based on my quite strongly negative opinion of the traditional Soviet science management system. Those who feel that the problems of the old system have been overstated in this book can presumably feel more optimistic about the outcome of Gorbachev's efforts. But my judgement is that for the foreseeable future Soviet science management will continue to be dysfunctionally centralised *de jure* and decentralised *de facto*, that the incentives driving researchers will continue not to be performance maximising, that management will continue to be authoritarian and therefore tend to be conservative, and that a majority of rank-and-file staff will continue to accept this situation as in their best interests. This means that Soviet science will continue to perform at a lower level than might be expected.

Notes

Chapter 1 Investment and return

1 L.E. Nolting, *Financing of Research, Development and Innovation in the USSR*, Foreign Economic Report No.22, Department of Commerce, Washington, DC, September 1985, p. 19; E. Zaleski et al., *Science Policy in the USSR*, OECD, Paris, 1969, pp. 95–6.
2 Nolting, *Financing of RDI*, p. 56; *Science Indicators. The 1985 report*, National Science Board, Washington, DC, 1985, p. 5.
3 Soviet data show their national income at about one-third the US level. *Narodnoe khoziaistvo SSSR za 70 let. 1917–87. lubileinyi statisticheskii ezhegodnik*, Finansy i statistika, Moscow, 1987, p. 13.
4 *Pravda* 19 June 1986, p. 2.
5 A.I. Shcherbakov, *Effektivnost' nauchoi deiatel'nosti v SSSR*, Ekonomika, Moscow, 1982, p. 30.
6 V.I. Nedil'ko, *Intensifikatsiia nauki: soderzhanie i planovo-finansovye faktory*, Nauka i tekhnika, Minsk, 1984, pp. 73–4.
7 *Finansy SSSR* no.10 (1981), p. 35.
8 *VAN* no.8 (1987), p. 18.
9 Nedil'ko, *Intensifikatsiia nauki*, pp. 73–4.
10 *Planirovanie, tekhnicheskii progress, effektivnost'*, Ekonomika, Moscow, 1985, p. 116.
11 Nolting, *Financing of RDI*, pp. 15–17.
12 V.N. Arkhangel'skii, *Territorial'noe upravlenie razvitiem nauki i tekhniki*, Ekonomika, Moscow, 1985, p. 51.
13 *Pravda* 19 June 1986, p. 2.
14 *VAN* no.11 (1986), p. 25.
15 *Izvestiia* 11 July 1988, p. 2.
16 *VAN* No.7 (1980), p. 36.
17 *Osnovy upravleniia naukoi. Ekonomicheskie problemy*, Nauka, Moscow, 1983, p. 141.
18 *Izvestiia* 9 March 1986, p. 2.
19 *VAN* no.11 (1986), p. 25.
20 *Partiinye organizatsii Urala i nauchno-tekhnicheskii progress*, Ural'skii gos. un-t, Sverdlovsk, 1985, p. 38; Arkhangel'skii, *Territorial'noe upravlenie*, p. 51.
21 Nedil'ko, *Intensifikatsiia nauki*, p. 66.

22 L.S. Bliakhman and M.S. Mintairov (eds), *Normativnye osnovy planirovaniia nauchno-tekhnicheskogo progressa*, Nauka, Leningrad, 1981, p. 22.
23 Arkhangel'skii, *Territorial'noe upravlenie*, pp. 48–9.
24 *Sotsialisticheskaia industriia* 7 April 1988, p. 1. See also Nedil'ko, *Intensifikatsiia nauki*, pp. 81, 86.
25 V.A. Trapeznikov (ed.), *Upravlenie razvitiem nauki i tekhniki*, Ekonomika, Moscow, 1980, pp. 109–10; *VAN* no.2 (1966), p. 38; V.I. Duzhenkov, *Problemy organizatsii nauki (Regional'nye aspekty)*, Nauka, Moscow, 1978, p. 166.
26 *VAN* no.11 (1984), p. 109.
27 *VAN* no.7 (1987), p. 49. See also Nedil'ko, *Intensifikatsiia nauki*, pp. 86–7.
28 Shcherbakov, *Effektivnost'*, p. 166.
29 *VAN* no.5 (1976), p. 39; no.8 (1977), p. 42.
30 *VAN* no.5 (1976), p. 39.
31 *VAN* no.3 (1978), p. 119.
32 *VAN* no.11 (1986), p. 25.
33 *Science Indicators. The 1985 report*, p. 5. For other work on this issue, see Zaleski, *Science Policy*, part 2 and part 5, appendix C; R. A. Lewis, *The Research and Development Effort in the Soviet Union. The manpower data, 1962–1969*, CREES Discussion Paper, University of Birmingham, October 1971; C.P. Ailes and F.W. Rushing, *The Science Race. Training and utilization of scientists and engineers, US and USSR*, Crane Russak, New York, 1982, chapter 7.
34 *Sotsialisticheskaia industriia* 10 February 1987, p. 2.
35 Duzhenkov, *Problemy organizatsii*, p. 43; V.A. Pokrovskii, *Uskorenie nauchno-tekhnicheskogo progressa. Organizatsiia i metody*, Ekonomika, Moscow, 1983, p. 187.
36 *Kommunist Estonii* No.12 (1982), p. 53; G.D. Komkov, B.V. Levshin, and L.K. Semenov, *Akademiia nauk SSSR. Kratkii istoricheskii ocherk*, 2nd edn, 2 vols, Nauka, Moscow, 1977, vol. 2, pp. 291, 373.
37 A.N. Kalimanov, *Organizatsiia i planirovanie nauchno-tekhnologicheskoi deiatel'nosti v tselliulozno-bumazhnoi promyshlennosti*, Lesnaia promyshlennost', Moscow, 1983, p. 15; *Izvestiia* 4 July 1984, p. 2.
38 G.A. Lakhtin, *Upravlenie v nauchnom uchrezhdenii*, Energoatomizdat, Moscow, 1983, p. 58.
39 *Ekonomicheskaia gazeta* no.12 (1987), pp. 14–15.
40 Ailes and Rushing, *The Science Race*, p. 135; *Science* 1 February 1980, p. 497; Shcherbakov, *Effektivnost'*, p. 30; B.I. Kozlov (ed.), *Organizatsiia i razvitie otraslevykh nauchno-issledovatel'skikh institutov Leningrada. 1917–1977*, Nauka, Leningrad, 1979, p. 93.
41 M.A. Gusakov (ed.), *Organizatsionnye formy sviazi nauki s proizvodstvom v promyshlennosti*, Nauka, Leningrad, 1980, p. 81; Shcherbakov, *Effektivnost'*, p. 30; Ailes and Rushing, *The Science Race*, p. 139.
42 Zaleski, *Science Policy*, p. 511; *Science Indicators. The 1985 report*, pp. 5–6.
43 *Zaria Vostoka* 5 June 1982, p. 4.
44 Shcherbakov, *Effektivnost'*, p. 30.
45 *Osnovy upravleniia naukoi*, pp. 73–4.

46 J. Irvine and B.R. Martin, 'Basic research in the East and West: a comparison of the scientific performance of high-energy accelerators', *Social Studies of Science*, vol. 15 (1985), pp. 293–341.
47 F. Narin and M.P. Carpenter, 'National publication and citation comparisons', *Journal of the American Society for Information Science*, vol. 26 (1975), pp. 80–93; J.D. Frame and F. Narin, 'The international distribution of biomedical publications', *Federation Proceedings*, vol. 36 (May 1977), pp. 1790–5; J.D. Frame and J.J. Baum, 'Cross-national information flows: examples taken from physics', *Journal of the American Society for Information Science*, vol. 29 (September 1978), pp. 247–52; F. Narin, J.D. Frame, and M.P. Carpenter, 'Highly cited Soviet papers: an exploratory investigation', *Social Studies of Science*, vol. 13 (1983), pp. 307–19; J.D. Frame, and M.P. Carpenter, 'International research collaboration', *Social Studies of Science*, vol. 9, No.4 (December 1979), p. 490; *Nature* 14 October 1982, p. 568. For some recent Soviet scientometric measurement of its performance , see *VAN* No.5 (1988), p. 79.
48 H. Balzer, *Soviet Research and Development. Information and insights from the third emigration*, final report to National Council for Soviet and East European Research, Harvard University, 1 July 1986. Further reports are to be issued in the future. The survey's data have been extensively used in R.F. Miller, 'The role of the Communist Party in Soviet research and development', *Soviet Studies*, vol. 37, No.1 (January 1985), pp. 31–59.
49 Balzer, *Soviet R&D*, p. 91.
50 Published in the UK as *Science in Chains. The crisis of science and scientists in the Soviet Union today*, Collins and Harvill, London, 1980, and in the USA as *Manipulated Science*, Doubleday, New York, 1979. The Russian language edition is *Upravliaemaia nauka*, Overseas Publications Interchange, London, 1978.
51 For example, Z. Medvedev, *Soviet Science*, Oxford University Press, Oxford, 1979; M.Ya. Azbel, *Refusenik. Trapped in the Soviet Union*, Hamish Hamilton, London, 1982; R. Berg, *Sukhovei. Vospominaniia genetika*, Chalidze, New York, 1983.
52 *New Scientist* 10 November 1983, p. 442.
53 C. Sinclair (ed.), *The Status of Soviet Civil Science*, Nijhoff, Dordrecht, 1987.
54 C. Kaysen, chairman, *Review of US–USSR Interacademy Exchanges and Relations*, National Academy of Sciences, Washington, DC, 1977. For commentary, see *Science* 27 October 1978, pp. 383–90; L.L. Lubrano, 'National and international politics in US–USSR cooperation', *Social Studies of Science*, vol. 11, no.4 (November 1981), pp. 451–80; C.P. Ailes and A.E. Pardee, *Cooperation in Science and Technology: an evaluation of the US–Soviet agreement*, Westview, Boulder, CO, 1986.
55 T. Gustafson, 'Why doesn't Soviet science do better than it does?', in L.L. Lubrano and S.G. Solomon (eds), *The Social Context of Soviet Science*, Westview, Boulder, CO, and Dawson, Folkestone, 1980, p. 32.
56 I. Wirszup, 'Soviet secondary school mathematics and science programs', in Sinclair, *The Status*, chapter 1.2.
57 *Izvestiia* 10 August 1988, p. 2.

58 R.L. Garwin, chairman, *A Review of Cooperation in Science and Technology between the US and USSR*, National Academy of Sciences, Washington, DC, 1977; Gustafson, 'Why doesn't', p. 60, n.4.
59 P. Hanson and K. Pavitt, *The Comparative Economics of Research, Development and Innovation in East and West: a survey*, Science Policy Research Unit, University of Sussex, Brighton, February 1987. To be published in J. Lesbourne and H. Sonnenschein (eds), *Fundamentals of Pure and Applied Economics*, Harvard Academic Publishers, forthcoming.
60 R. Amann and J. Cooper (eds), *Technical Progress and Soviet Economic Development*, Blackwell, Oxford, 1986, pp. 9–10. Soviet data released since Amann's chapter show continuing declines in many of these indicators. *Vestnik statistiki* no.7 (1988), pp. 58–64.
61 Amann and Cooper, *Technical Progress*, pp. 8–9; Hanson and Pavitt, *The Comparative Economics of RDI*, para. 4.4.
62 R. Amann, J. Cooper, and R.W. Davies (eds), *The Technological Level of Soviet Industry*, Yale University Press, New Haven, CN, and London, 1977; R. Amann and J. Cooper (eds), *Industrial Innovation in the Soviet Union*, Yale University Press, New Haven, CN, and London, 1982; Amann and Cooper, *Technical Progress.*
63 R.W. Campbell, *The Economics of Oil and Gas*, Johns Hopkins University Press, Baltimore, MD, 1968; C. Rajana, *The Chemical and Petro-Chemical Industries of Russia and Eastern Europe, 1960–1980*, Sussex University Press, 1975; S.E. Goodman and W.K. McHenry, 'Computing in the Soviet general economy: an introductory overview', *Science and Technology in the Soviet Union. Proceedings of a conference*, Department of Statistics, Stanford University, January 1985; W.J. Kelly, H.L. Shaffer, and J.K. Thompson, *Energy Research and Development in the USSR*, Duke University Press, Durham, NC, 1986.
64 See Ronald Amann's comments in Amann and Cooper, *Technical Progress*, pp. 2–3.
65 *Ekonomika Sovetskoi Ukrainy* no.9 (1985), p. 72.
66 See, for example, *Sotsialisticheskaia industriia* 14 January 1986, p. 3.
67 For a summary of Soviet views, see L.E. Nolting, *The Soviet View of the State of Technological Innovation in the USSR*, Center for International Research Staff Paper no. 38, US Bureau of the Census, Washington, DC, April 1988.
68 For examples, see *Sotsialisticheskaia industriia* 10 April 1986, p. 2; 21 May 1986, p. 2; *Izvestiia* 9 January 1986, p. 2.

Chapter 2 Central party and governmental agencies

1 S. Fortescue, *The Communist Party and Soviet Science*, Macmillan, London and Basingstoke, 1986, pp. 57–9.
2 *VAN* no.11 (1986), p. 9.
3 S. Freidzon, *Top-Level Administration of the Soviet Economy: a partial view*, Rand Paper P-7178, Santa Monica, CA, January 1986, p. 102.
4 *Kommunist* no.12 (1979), pp. 34–5.
5 *Khrushchev Remembers. The last testament*, Deutsch, London, 1974, p. 60.
6 For example, *Physics Today* July 1988, p. 50.

7 L.I. Brezhnev, *Tselina*, Politizdat, Moscow, 1978, pp. 14, 17.
8 *Khrushchev Remembers*, pp. 63–8.
9 *Khrushchev Remembers*, p. 62.
10 *Khrushchev Remembers*, p. 64.
11 E. Andronikashvili, *Nachinaiu s El'brusa (Tvorcheskie portrety uchenykh)*, Metsnieraba, Tbilisi, 1982, pp. 58–62.
12 *Khrushchev Remembers*, p. 63.
13 *Leningradskaia pravda* 17 August 1985, p. 2.
14 *VAN* no.3 (1981), p. 84.
15 Decrees have appeared in April 1976, July 1981, and August 1985. Politburo discussions have been reported in May and August 1983 and August 1985.
16 *Partiinaia zhizn'* no.19 (1984), p. 3; *Leningradskaia pravda* 1 March 1985, p. 1; *Izvestiia* 28 April 1986, p. 2; *Sotsialisticheskaia industriia* 9 May 1986, p. 1. It appears that the pressure is considered to have paid off, since rotor technology is now rated in the Soviet Union as one of its few recent S&T successes. *Izvestiia* 25 August 1987, p. 2.
17 Fortescue, *The Communist Party*, p. 74.
18 *Pravda* 29 November 1988, p. 2.
19 Dubinin was one of the protagonists in a rather nasty debate over the relative influence of 'nature' or 'nurture' in human development. He lost the debate and at the same time the directorship of his institute and his position as advisor to the Central Committee apparatus. Fortescue, *The Communist Party*, pp. 24–5; R. Berg, *Sukhovei. Vospominaniia genetika*, Chalidze, New York, 1983, p. 249; L.R. Graham, *Science, Philosophy and Human Behavior in the Soviet Union*, Columbia University Press, New York, 1987, chapter 6.
20 M. Voslensky, *Nomenklatura. Anatomy of the Soviet ruling class*, Bodley Head, London, 1984, chapter 10, section 9.
21 For detailed accounts, see M. Azbel, *Refusenik. Trapped in the Soviet Union*, Hamish Hamilton, London, 1982; Z. Medvedev, *The Medvedev Papers*, Macmillan, London, 1971; M. Popovsky, *Science in Chains. The crisis of science and scientists in the Soviet Union today*, Collins and Harvill, London, 1980.
22 *Radio Liberty Research Bulletin* RL Supplement 5/88, 1 November 1988.
23 For more details on them, see Freidzon, *Top-Level Administration*, esp. chapters 11 and 12.
24 Freidzon, *Top-Level Adminstration*, p. 149.
25 In his keynote speech to the plenum Kosygin had briefly foreshadowed the change. *Pravda* 28 September 1965, p. 4.
26 For a list of predecessors, see E. Zaleski et al., *Science Policy in the USSR*, OECD, Paris, 1969, pp. 52–7.
27 *Pravda* 3 October 1965, p. 1.
28 *SP SSSR* no.21 (1966), art.193, para.2.
29 *VAN* no.8 (1976), p. 33.
30 V.A. Pokrovskii, *Uskorenie nauchno-tekhnicheskogo progressa. Organizatsiia i metody*, Ekonomika, Moscow, 1983, p. 126.
31 L.R. Graham, 'The role of the Academy of Sciences', *Survey*, vol.23, no.1 (102) (Winter 1977–8), p. 119.

32 *Sotsialisticheskaia industriia* 28 August 1983, p. 1; N.A. Gordeeva, *Vzaimosviazi akademicheskoi nauki i proizvodstva. Organizatsionno-pravovoi aspekt*, Nauka, Moscow, 1985, p. 177.
33 S.I. Aivazian, Iu.A. Vedeneev, and O.A. Supataeva, *Pravovye voprosy upravleniia nauchnymi issledovaniiami*, Nauka, Moscow, 1980, p. 64.
34 Graham, 'The role of the Academy', p. 119. See also *Pravda* 18 August 1970, p. 3.
35 For an example, see Andronikashvili, *Nachinaiu*, pp. 276–8. See also Iu.M. Kanygin and V.A. Botvin, *Problemy razvitiia i ispol'zovaniia nauchnogo potentsiala krupnykh gorodov*, Naukova dumka, Kiev, 1980, p. 183.
36 *Sotsial'nye i ekonomicheskie problemy povysheniia effektivnosti nauki*, Nauka, Moscow, 1985, p. 148.
37 *Voprosy ekonomiki* no.11 (1981), p. 156; *Pravda* 8 July 1982, p. 2.
38 *Izvestiia* 20 November 1974, p. 1.
39 *Gosudarstvennoe upravlenie nauchno-tekhnicheskim progressom v promyshlennosti soiuznoi respubliki (na materialakh Kazakhskoi SSR)*, Nauka, Alma-Ata, 1984, p. 100.
40 Aivazian, *Pravovye voprosy*, p. 64.
41 *Sobranie postanovlenii SSSR* no.21 (1966), art.193, para.4j.
42 *Pravda* 21 July 1982, p. 2; *Kommunist* no.4 (1983), p. 68; *Izvestiia* 3 July 1984, p. 2.
43 *Pravda* 9 April 1987, p. 3; *Moscow News* 5 July 1987, p. 10.
44 For some examples, see *Sotsialisticheskaia industriia* 19 April 1973, p. 1; 20 March 1974, p. 2; 14 May 1976, p. 2; 5 March 1985, p. 2; 5 March 1986, p. 4; 4 September 1985, p. 1; *Pravda* 27 December 1985, p. 2; 8 January 1986, p. 3; *VAN* no.11 (1981), pp. 55–6; *Trud* 28 November 1967, p. 2.
45 *Sotsialisticheskaia industriia* 29 January 1986, p. 2.
46 *Literaturnaia gazeta* 26 March 1975, p. 10; *Sotsialisticheskaia industriia* 11 August 1973, p. 2.
47 *Sovetskaia Rossiia* 27 May 1983, p. 2.
48 For details on GKNT's structure, see CIA Scientific Institutions, *USSR State Committee for Science and Technology. A reference aid*, CR81-11110, Central Intelligence Agency, Washington, DC, June 1981.
49 *Sotsialisticheskaia industriia* 17 July 1985, p. 2.
50 *Sotsialisticheskaia industriia* 27 November 1977, p. 2.
51 *Pravda* 1 October 1985, p. 3. For other examples, see *Pravda* 3 August 1985, p. 2; 11 February 1980, p. 3; 11 August 1985, p. 3; *Sotsialisticheskaia industriia* 14 October 1986, p. 2.
52 Aivazian, *Pravovye voprosy*, p. 77; R. Amann and J. Cooper (eds), *Industrial Innovation in Soviet Union*, Yale University Press, New Haven, CN, and London, 1982, pp. 488–9; *Planovoe khoziaistvo* no.6 (1978), p. 99.
53 *Kommunist* no.4 (1983), p. 71.
54 *VAN* no.11 (1979), pp. 43–53; no.11 (1981), pp. 53–61; *Sovetskoe gosudarstvo i pravo* no.3 (1985), pp. 67–72. For references to similar calls by other authors, see Graham, 'The role of the Academy', p. 120.
55 *Leningradskaia pravda* 12 June 1985, p. 2; *Pravda* 9 November 1985, p. 2.
56 *Izvestiia* 11 March 1986, p. 2.
57 *Izvestiia* 19 August 1986, p. 2.

58 *Izvestiia* 20 September 1986, p. 2; *Ekonomicheskaia gazeta* no.11 (1986), p. 26.
59 For his biography, see *Izvestiia* 8 February 1987, p. 1.
60 *Pravda* 1 May 1987, p. 1.
61 *O korennoi perestroike upravleniia ekonomikoi. Sbornik dokumentov*, Politizdat, Moscow, 1987, p. 94.
62 *Kommunist* no.1 (1987), p. 31.
63 *Izvestiia* 7 September 1986, p. 2; 3 October 1986, p. 2.
64 *VAN* no.1 (1987), p. 66.

Chapter 3 The Academy of Sciences

1 In 1917 the Academy had 154 scientific workers. *Narodnoe khoziaistvo SSSR za 70 let. 1917–1987. Iubileinyi statisticheskii ezhegodnik*, Finansy i statistika, Moscow, 1987, p. 60. For a good summary of the pre-revolutionary history of the Academy, see A. Vucinich, *Empire of Knowledge. The Academy of Sciences of the USSR (1917–1970)*, University of California Press, Berkeley and Los Angeles, CA, 1984, chapter 1.
2 L.R. Graham, *The Soviet Academy of Sciences and the Communist Party, 1927–1932*, Princeton University Press, Princeton, NJ, 1967.
3 *Pravda* 11 and 12 March 1987, p. 3.
4 See the role of Suslov in the election of Aleksandrov and of Ligachev in the election of Marchuk.
5 S. Fortescue, 'The Soviet Academy of Sciences under Gorbachev', *Australian Slavonic and East European Studies* vol. 1, no.2 (1987), pp. 4–8.
6 *VAN* no.7 (1986), p. 52.
7 Vucinich, *Empire of Knowledge*, p. 309.
8 Such decisions are printed at the back of the Academy's journal *Vestnik Akademii nauk SSSR*.
9 *Vospominaniia ob akademike L.A. Artsimoviche*, Nauka, Moscow, 1981, p. 73; *VAN* no.7 (1976), p. 86.
10 *Vospominaniia ob Artsimoviche*, pp. 73, 82–3.
11 *VAN* no.9 (1976), p. 84.
12 *VAN* no.10 (1987), p. 135.
13 For a good account of the debate, see L.L. Greenberg, 'Policy-making in the USSR Academy of Sciences', *Journal of Contemporary History*, vol.8, no.4 (October 1973), pp. 67–80.
14 *VAN* no.7 (1986), pp. 74–7.
15 *VAN* no.11 (1986), pp. 27–8; no.1 (1987), p. 53.
16 *VAN* no.11 (1986), pp. 20–1; *Pravda* 16 February 1987, p. 3.
17 *VAN* no.1 (1987), pp. 6–9; *Pravda* 13 March 1987, p. 2; *Izvestiia* 22 March 1987, p. 2.
18 *VAN* No.11 (1986), p. 12; *Pravda* 18 December 1986, p. 2; *Sotsialisticheskaia industriia* 17 March 1987, p. 1.
19 For a 1963 list of such specialisations, see *VAN* no.6 (1963), pp. 7–8.
20 *Pravda* 9 July 1986, p. 3; *VAN* no.1 (1987), p. 11.
21 *VAN* no.1 (1987), p. 11; no.5 (1987), pp. 3–5; *Izvestiia* 22 March 1987, p. 2.
22 *Pravda* 24 December 1987, p. 3.
23 *Izvestiia* 14 September 1984.

24 *VAN* no.7 (1981), p. 6.
25 Z.A. Medvedev, *Soviet Science*, Oxford University Press, Oxford, 1979, p. 188. See also *The Guardian* 19 June 1975, p. 16.
26 *Vospominaniia o I.E. Tamme*, 2nd edn, Nauka, Moscow, 1986, p. 124.
27 *The Guardian* 19 June 1975, p. 16; Iu. Glazov, *Tesnye vrata. Vozrozhdenie russkoi intelligentsii*, Overseas Publications Interchange, London, 1973, pp. 128–9.
28 M. Voslensky, *Nomenklatura. Anatomy of the Soviet ruling class*, Bodley Head, London, 1984, p. 358. For other cases, see A. Sakharov, 'Fragmenty iz avtobiograficheskoi knigi', *Kontinent* no.36 (1983), pp. 151–60; Z.A. Medvedev, *The Rise and Fall of T.D. Lysenko*, Columbia University Press, New York and London, 1969, pp. 214–17.
29 *Ogonek* no.4 (1988), p. 29.
30 For a summary of these events see *Ogonek* no.4 (1988), pp. 28–9.
31 *Izvestiia* 24 December 1987, p. 3; 11 February 1988, p. 5.
32 *Ogonek* no.4 (1988), p. 29.
33 *The Guardian* 19 June 1975, p. 16.
34 *Times Higher Educational Supplement* 26 March 1982, p. 12.
35 For more details, see S. Fortescue, 'Party membership in Soviet research institutes', *Soviet Union/Union Sovietique* vol.11, pt 2 (1984), pp. 129–38.
36 For more detailed discussion of the Academy involvement in defence R&D, see J.R. Thomas, 'Militarization of the Soviet Academy of Sciences', *Survey*, vol.29, no.1 (124) (Spring, 1985), pp. 29–58.
37 *VAN* no.11 (1986), p. 30.
38 *Izvestiia* 11 February 1988, p. 5. See also *VAN* no.5 (1988), p. 12.
39 *VAN* no.2 (1986), pp. 36–7.
40 *VAN* no.10 (1987), p. 38. New data on numbers of research institutes in the science sector as a whole have appeared in recent statistical publications, but numbers of Academy institutes are not included.
41 *VAN* no.6 (1983), pp. 11–13, 41.
42 *VAN* no.6 (1977), pp. 8–9, 69–70.
43 *VAN* no.7 (1987), p. 29.
44 *Izvestiia* 22 March 1987, p. 2.
45 *VAN* no.8 (1987), p. 9.
46 *VAN* no.5 (1986), p. 8.
47 L.E. Nolting, *Financing of Research, Development and Innovation in the USSR*, Foreign Economic Report No.22, Department of Commerce, Washington, DC, September 1985, p. 140.
48 *Ekonomicheskoe upravlenie naukoi (mezhotraslevye voprosy)*, proceedings of symposium 'Sotsial'nye i ekonomicheskie aspekty povysheniia effektivnosti sovetskoi nauki', Zvenigorod, 24–26 May 1982, p. 72; M.J. Berry (ed.), *Science and Technology in the USSR*, Longman, Harlow, 1988, p. 385.
49 *VAN* no.1 (1986), p. 23.
50 *EKO* no.5 (1986), pp. 59–61.
51 *Pravda* 12 June 1985, p. 2.
52 Berry, *Science and Technology*, p. 385.
53 *VAN* no.11 (1986), p. 25.
54 *Izvestiia* 17 October 1985, p. 2; *Pravda* 9 July 1986, p. 3; *VAN* no.3 (1986),

p. 39; no.5 (1986), p. 41; no.1 (1987), pp. 39, 52.
55 *VAN* no.11 (1986), pp. 20–1.
56 *VAN* no.1 (1987), p.3.
57 For a brief account of the allocations to high-energy physics, see Fortescue, 'The Soviet Academy of Sciences', pp. 17–19. On developments in engineering science, see *Pravda* 20 March 1986, p. 3.
58 *Pravda* 25 February 1976, p. 6.
59 *VAN* no.6 (1977), p. 69.
60 N.A. Gordeeva, *Vzaimosviazi akademicheskoi nauki i proizvodstva. Organizatsionno-pravovoi aspekt*, Nauka, Moscow, 1985, p. 128.
61 *VAN* no.7 (1986), p. 76.
62 *VAN* no.1 (1986), p. 7
63 *VAN* no.7 (1986), p. 77; Gordeeva, *Vzaimosviazi*, p. 128.
64 For a brief historical account of the councils, see Vucinich, *Empire of Knowledge*, pp. 310–13.
65 *VAN* no.5 (1976), p. 17; no.9 (1976), pp. 19, 75; no.1 (1981), p. 68.
66 S.I. Aivazian, Iu.A. Vedeneev, and O.A. Supataeva, *Pravovye voprosy upravleniia nauchnymi issledovaniiami*, Nauka, Moscow, 1980, p. 70.
67 *VAN* no.3 (1981), pp. 65–70.
68 *VAN* no.1 (1988), pp. 57–9.
69 S. Fortescue, 'Project planning in Soviet R&D', *Research Policy* vol.14, (1985), pp. 270–1.
70 Aivazian, *Pravovye voprosy*, pp. 158–61.
71 Gordeeva, *Vzaimosviazi*, p. 42.
72 *VAN* no.5 (1972), pp. 19–20.
73 *Pravda* 4 March 1982, p. 2.
74 R. Amann and J. Cooper (eds), *Industrial Innovation in the Soviet Union*, Yale University Press, New Haven, CN, and London, 1982, p. 471; Gordeeva, *Vzaimosviazi*, pp. 55–9; *Izvestiia* 30 January 1985, p. 2; *Pravda* 7 April 1987, p. 3.
75 *Literaturnaia gazeta* 17 February 1988, p. 10; *Kommunist* no.17 (1987), p. 60.
76 *VAN* no.4 (1976), pp. 10–11; no.11 (1983), pp. 121–4; *Sotsialisticheskaia industriia* 5 February 1980, p. 2.
77 *VAN* no.3 (1983), p. 127.
78 See S. Fortescue, *The Communist Party and Soviet Science*, Macmillan, London and Basingstoke, 1986, pp. 107–8.
79 *Pravda* 13 December 1985, p. 1; *Izvestiia* 23 January 1986, p. 2.
80 S. Fortescue, *The Academy Reorganized. The R&D role of the Soviet Academy of Sciences since 1961*, Occasional Paper No.17, Department of Political Science, Research School of Social Sciences, Australian National University, Canberra, 1983, p. 33.

Chapter 4 The branch research and development network

1 For a summary of these problems, see S. Fortescue, 'Recent developments in the adminstration of branch ministry research', in C. Sinclair (ed.), *The Status of Soviet Civil Science*, Nijhoff, Dordrecht, 1987, chapter 2.3.
2 J. Richardson, 'The Soviet ministerial élite in the Brezhnev era', seminar

paper, Department of Political Science, Research School of Social Sciences, Australian National University, Canberra, 7 June 1982; J.F. Hough, *The Soviet Prefects. The local party organs in industrial decision-making*, Harvard UP, Cambridge, MA, 1969, p.48.

3 E.A. Beliaev, *KPSS i organizatsiia nauki SSSR*, Politizdat, Moscow, 1982, p. 103; B.V. Aver'ianov, *Funktsii i organizatsionnaia struktura organa gosudarstvennogo upravleniia*, Naukova dumka, Kiev, 1979, p. 79.

4 I.L. Davitnidze, *Kollegii ministerstv (Pravovoe polozhenie i organizatsiia raboty)*, Iuridizdat, Moscow, 1972, pp. 60–1.

5 Davitnidze, *Kollegii*, p. 63; V.S. Pronina, *Tsentral'nye organy upravleniia narodnym khoziaistvom*, Iuridizdat, Moscow, 1971, p. 81.

6 N.G. Kalinin (ed.), *Organizatsiia upravleniia v sisteme ministerstv*, Izd-vo Moskovskogo un-ta, Moscow, 1974, p. 65; Davitnidze, *Kollegii*, pp. 64–9; Pronina, *Tsentral'nye organy*, p. 83; *Pravda* 10 March 1969, p. 2; *Sotsialisticheskaia industriia* 23 January 1986, p. 2; *Gosudarstvennoe upravlenie nauchno-tekhnicheskim progressom v promyshlennosti soiuznoi respubliki (na materialakh Kazakhskoi SSR)*, Nauka, Alma-Ata, 1984, p. 106.

7 Davitnidze, *Kollegii*, p. 94.

8 *Izvestiia* 4 September 1985, p. 2.

9 *Pribory i sistemy upravleniia* no.10 (1985), p. 2.

10 *Sotsialisticheskaia industriia* 14 June 1986, p. 2.

11 K. Dubrovskii, *Otsenka i upravlenie nauchnymi issledovaniiami v otrasli promyshlennosti*, presentation to an international symposium of scientists and specialists of CMEA countries and Yugoslavia, 'NTP i sotsial'nyi progress', section 5, 'Problemy otsenki effektivnosti nauchno-tekhnicheskikh razrabotok i novoi tekhniki', Moscow, 1974, p. 222.

12 Davitnidze, *Kollegii*, p. 146.

13 The *tipovoe polozhenie* (Model Statute) for the STC states that either the minister or his first deputy can chair the body. *Sobranie postanovlenii SSSR* no.17 (1966), art.154, para.4.

14 *SP SSSR* no.17 (1966), art.154, para.4.

15 *Trud* 5 March 1969, p.2.

16 *SP SSSR* no.17 (1966), art.154, para.5.

17 Kalinin, *Organizatsiia upravleniia*, p. 66.

18 N.G. Kalinin, *General'nye skhemy upravleniia*, Znanie, Moscow, 1982, p. 39; *SP SSSR* no.17 (1966), art.154, para.7.

19 For a rather wordier list of the body's responsibilities, see *SP SSSR* no.17 (1966), art.154, paras 2 and 3.

20 *Sotsialisticheskaia industriia* 4 January 1986, p. 2.

21 *Standarty i kachestvo* no.4 (1977), p. 54; *Gosudarstvennoe upravlenie*, p. 54.

22 S.L. Vorotenitskaia, *Upravlenie kachestvom produktsii v pishchevoi promyshlennosti*, Izd-vo standartov, Moscow, 1983, p. 40.

23 P.M. Cocks, *Science Policy USA/USSR*. vol.2, *Science policy in the USSR*, National Science Foundation, Washington, DC, 1980, pp. 146–7.

24 *Pravda* 2 April 1985, p. 2.

25 S.I. Aivazian, Iu.A. Vedeneev, and O.A. Supataeva, *Pravovye voprosy*

upravleniia nauchnymi issledovaniiami, Nauka, Moscow, 1980, p. 59.
26 Pronina, *Tsentral'nye organy*, p. 99.
27 V.G. Vishniakov, *Struktura i shtaby organov sovetskogo gosudarstvennogo upravleniia*, Nauka, Moscow, 1972, pp. 71–2.
28 Pronina, *Tsentral'nye organy*, p. 99; Aivazian, *Pravovye voprosy*, p. 29.
29 D.M. Gvishiani (ed.), *Upravlenie, planirovanie i organizatsiia nauchnykh i tekhnicheskikh issledovanii*, VINITI, Moscow, 1970, vol.4, p. 116; R.A. Lewis, 'Hierarchy and technological innovation in Soviet industry: the science–production associations', *Minerva* vol.22, no.2 (Summer 1984), p. 148.
30 Aivazian, *Pravovye voprosy*, p. 49; *Izvestiia* 3 July 1984, p. 2; *Standarty i kachestvo* no.4 (1977), p. 54.
31 V.S. Sominskii, V.I. Kublis, and K.G. Fedorov, *Upravlenie naukoi v khimicheskoi promyshlennosti*, Khimiia, Moscow, 1978, p. 18; *Gosudarstvennoe upravlenie*, p. 100.
32 Aver'ianov, *Funktsii*, p. 79; Kalinin, *Organizatsiia upravleniia*, pp. 71–4.
33 For example, *Sotsialisticheskaia industriia* 1 August 1985, p. 2.
34 Kalinin, *Organizatsiia upravleniia*, pp. 101–2.
35 *Gosudarstvennoe upravlenie*, p. 100; L.E. Nolting, *Financing of Research, Development and Innovation in the USSR*, Foreign Economic Report No.22, Department of Commerce, Washington, DC, September 1985, p. 32.
36 N.A. Gordeeva, *Vzaimosviazi akademicheskoi nauki i proizvodstva. Organizatsionno-pravovoi aspekt*, Nauka, Moscow, 1985, pp. 136–7; *Gosudarstvennoe upravlenie*, p. 100.
37 *Sotsialisticheskaia industriia* 20 July 1985, p. 2; *Izvestiia* 4 July 1984, p. 2; Sominskii, *Upravlenie naukoi*, pp. 17, 161.
38 Vorotenitskaia, *Upravlenie kachestvom*, pp. 30–4.
39 *Partiinaia zhizn'* no.1 (1986), p. 48.
40 Vorotenitskaia, *Upravlenie kachestvom*, pp. 30–4.
41 *Ekonomicheskaia gazeta* no.47 (1985), p. 6; *Sotsialisticheskaia industriia* 15 November 1985, p. 1.
42 Kalinin, *General'nye skhemy*, p. 39.
43 Sominskii, *Upravlenie naukoi*, p. 18.
44 *Sotsialisticheskaia industriia* 29 June 1985, p. 2; 10 August 1985, p. 2; *Pravda* 16 July 1985, p. 2; 11 August 1985, p. 3; *Sovetskaia Rossiia* 1 November 1985, p. 3; *Voprosy istorii KPSS* no.4 (1984), p. 93.
45 *Sotsialisticheskaia industriia* 4 January 1986, p. 2.
46 *Pravda* 11 August 1985, p. 2. See also *Sovetskaia Rossiia* 1 November 1985, p. 3.
47 *Pravda* 11 August 1985, p. 3.
48 *Izvestiia* 3 July 1984, p. 2.
49 *Sotsialisticheskaia industriia* 29 June 1985, p. 2; *Pravda* 16 July 1985, p. 2.
50 R. Amann and J. Cooper (eds), *Industrial Innovation in the Soviet Union*, Yale University Press, New Haven, CN, and London, 1982, p. 94. For criticisms of other head institutes, see V.A. Pokrovskii, *Uskorenie nauchno-tekhnicheskogo progressa. Organizatsiia i metody*, Ekonomika, Moscow, 1983, pp. 90–3; *Pravda* 6 October 1973, p. 2; *EKO* no.2 (1984), pp. 55–6; *Sovetskaia Rossiia* 11 March 1983, p. 2; 25 February 1985, p. 2; *Izvestiia* 30

January 1985, p. 2; *Sotsialisticheskaia industriia* 21 December 1985, p. 2.
51 Kalinin, *Organizatsiia upravleniia*, p. 68.
52 *Sotsialisticheskaia industriia* 5 April 1985, p. 2.
53 B.Z. Mil'ner (ed.), *Sovershenstvovanie upravleniia proizvodstvom novoi tekhniki*, Nauka, Moscow, 1984, p. 45.
54 Mil'ner, *Sovershenstvovanie upravleniia*, p. 45; *Pravda* 4 November 1973, p. 2.
55 *Sotsialisticheskaia industriia* 19 July 1985, p. 2.
56 For a good example, see *Izvestiia* 16 January 1986, p. 6.
57 This is an issue of enormous fluidity at the moment, but some sense of the factors involved can be gained from S. Fortescue, 'The regional party apparatus in the "sectional society"', *Studies in Comparative Communism* vol.21, no.1 (Spring 1988), pp. 11–23.
58 *Pravda* 6 July 1977, p. 3; *Sovetskoe gosudarstvo i pravo* no.1 (1977), pp. 28–36; no.6 (1982), pp. 38–48; *Kommunist* no.18 (1982), pp. 48–59; *EKO* no.10 (1983), pp. 34–57.
59 At the June 1985 Central Committee conference and in Dnepropetrovsk in July 1985. *Pravda* 12 June 1985, p. 2; *Ekonomicheskaia gazeta* no.27 (1985), p. 6.
60 *Pravda* 18 October 1985, p. 1.
61 *Izvestiia* 11 March 1986, p. 2.
62 *Izvestiia* 19 August 1986, p. 2.
63 *Sovetskaia Rossiia* 29 May 1986, p. 2.
64 *Izvestiia* 3 May 1986, p. 2; *Voprosy istorii KPSS* no.3 (1988), p. 149; *Sotsialisticheskaia industriia* 8 August 1987, p. 2.
65 *Sotsialisticheskaia industriia* 20 November 1987, p. 1.
66 *Izvestiia* 19 August 1986, p. 2.
67 For Konyshev's details, see *Sotsialisticheskaia industriia* 5 April 1987, p. 3.
68 *Ekonomicheskaia gazeta* no.11 (1988), p. 18.
69 *Izvestiia* 4 August 1987, p. 2.

Chapter 5 Defence, *vuz*, and factory research and development

1 The main sources used here are D. Holloway, 'Innovation in the defence sector', in R. Amann and J. Cooper (eds), *Industrial Innovation in the Soviet Union*, Yale University Press, New Haven, CN, and London, 1982, chapter 7; J. Cooper, 'The Soviet Union', in S. McLean (ed.), *How Nuclear Decisions are Made*, Macmillan, London and Basingstoke, 1986, chapter 1; J. Cooper, 'Civilian and military engineering', in M.J. Berry, *Science and Technology in the USSR*, Longman, Harlow, 1988, chapter 22; S. Woods, 'Weapons acquisition in the Soviet Union', *Soviet Union/Union Sovietique* vol.10, pts 2–3 (1983), pp. 337–73; H. Balzer, *Soviet Research and Development. Information and insights from the third emigration*, final report to the National Council for Soviet and East European Research, Harvard University, 1 July 1986, chapter 4; A.J. Alexander, 'Soviet weapons development and the scientific community', in *Science and Technology in the Soviet Union. Proceedings of a conference*, Department of Statistics, Stanford University, January 1985, chapter 2; M. Agursky, *The Research Institute of Machine-Building Technology. A part of the Soviet*

military–industrial complex, Soviet Institutions Series Paper No.8, The Hebrew University of Jerusalem, September 1976.

2 Holloway, 'Innovation', pp. 354–67.

3 Cooper, 'The Soviet Union', p. 14.

4 Cooper, 'Civilian and military', p. 286. Some of the less sensitive institutes in the defence industry ministries are listed in this source.

5 J. Cooper, 'The civilian production of the Soviet defence industry', in R. Amann and J. Cooper (eds), *Technical Progress and Soviet Economic Development*, Blackwell, Oxford, 1986, chapter 2. For a recent updating of his data, see J. Cooper, *The Scales of Output of Civilian Products by Enterprises of the Soviet Defence Industry*, CREES Discussion Paper, Soviet Industry and Technology Series No.3, Centre for Russian and East European Studies, University of Birmingham, August 1988.

6 Balzer, *Soviet R&D*, p. 101.

7 According to *émigré* accounts reported by Arthur J. Alexander about half of Academy institutes have contributed to military research. Alexander, 'Soviet weapons', p. 32.

8 Alexander, 'Soviet weapons', p. 33.

9 For a recent summary of the debate, see J. Tedstrom, 'The economics of Soviet defense spending', *Radio Liberty Research*, RL Supplement 4/88 (15 August 1988).

10 J.F. Hough, 'The historical legacy in Soviet weapons development', in J. Valenta and W.C. Potter (eds), *Soviet Decision-making for National Security*, Allen and Unwin, London, 1984, p. 95.

11 R.W. Campbell, 'Management spillovers from Soviet space and military programmes', *Soviet Studies* vol.23, no.4 (April 1972), pp. 586–607.

12 Balzer, *Soviet R&D*, p. 99.

13 See the press report of a recent conference of military R&D managers. *Krasnaia zvezda* 14 August 1988.

14 Tedstrom, 'The economics', p. 13.

15 Balzer, *Soviet R&D*, pp. 115–18.

16 J.R. Sutton, 'Organizational autonomy and professional norms in science: a case study of the Lawrence Livermore Laboratory', *Social Studies of Science* vol.14, no.2 (May 1984), pp. 197–224.

17 J. Cooper, 'Western technology and the Soviet defense industry', in B. Parrott (ed.), *Trade, Technology, and Soviet–American Relations*, Indiana University Press, Bloomington, IN, 1985, p. 188.

18 This was most obvious in Khrushchev's relationship with Kapitsa. *Khrushchev Remembers. The last testament*, Deutsch, London, 1974, pp. 63–8.

19 *Akademik I.I. Artobolevskii*, Znanie, Moscow, 1983, pp. 29–30.

20 D. Holloway, *The Soviet Union and the Arms Race*, Yale University Press, New Haven, CN, and London, 1983, p. 23. It is interesting that recently the contribution of Klaus Fuchs has been recognised in Soviet sources, even if not to the degree that they are prepared to admit that the bomb was not all their own work. *Moscow News* 17 July 1988, p. 11; 20 December 1987, p. 7; *Izvestiia* 23 July 1988, p. 3.

21 COCOM is the committee of Western states which draws up and attempts to enforce bans on sales of high-technology goods to the Socialist bloc.

22 *Pacific Defence Reporter* July 1987, p. 48; August 1987, pp. 38–9; September 1987, pp. 23–4. For a Soviet account, see *Moscow News* no.31 (1987), p. 5.
23 I have based this section on the unpublished paper by Philip Hanson, *Soviet Industrial Espionage: some new information*. CREES, University of Birmingham, 1986. The most detailed published source is Thierry Wolton, *Le KGB en France*, Grasset, Paris, 1986.
24 *Soviet Acquisition of Militarily Significant Western Technology. An update*, USGPO, Washington, DC, 1985.
25 There is evidence that GKNT might run an independent network for civilian technology, while the Pentagon study also refers to the separate efforts of the Ministry of Foreign Trade. The exclusion of the Ministry of Foreign Trade from the VPK's acquisition agencies in the Farewell Documents is striking.
26 For more on the espionage involvement of exchange scientists, see Yu.M. Rabkin, 'Soviet organization of international scientific contacts', in C. Sinclair (ed.), *The Status of Soviet Civil Science*, Nijhoff, Dordrecht, 1987, chapter 2.1.
27 Cooper, 'Western technology'.
28 For details, see H. Wienert, 'The organisation and planning of research in the higher educational establishments', in E. Zaleski et al. (eds), *Science Policy in the USSR*, OECD, Paris, 1969, part 4; M.J. Berry (ed.), *Science and Technology in the USSR*, Longman, Harlow, 1988, chapter 6.
29 *Pravda* 9 March 1988, p. 2.
30 Berry, *Science and Technology*, pp. 65–6.
31 See a provincial *vuz* rector regretting the lack in his region of an Academy institution, and therefore of a focus for high-quality research, in *Sovetskaia Rossiia* 4 March 1979, p. 2.
32 Berry, *Science and Technology*, p. 64.
33 Berry, *Science and Technology*, p. 64; *VAN* no.8 (1987), p. 72.
34 A. Kukrus, *Nauka i upravlenie (upravlenie naukoi v SSSR)*, Valgus, Tallin, 1982, p. 45.
35 *Sotsialisticheskaia industriia* 13 November 1987, p. 1; Berry, *Science and Technology*, p. 64.
36 Berry, *Science and Technology*, p. 64.
37 Berry, *Science and Technology*, p. 67; *Sotsialisticheskaia industriia* 13 November 1987, p. 1.
38 *Pravda* 7 January 1987, p. 1.
39 For details, see Berry, *Science and Technology*, pp. 67–9.
40 C.P. Ailes and F.W. Rushing, *The Science Race. Training and utilization of scientists and engineers, US and USSR*, Crane Russak, New York, 1982, p. 139; A.V. Bakunin (ed.), *Deiatel'nost' KPSS po uskoreniiu nauchno-tekhnicheskogo progressa*, Vysshaia shkola, Moscow, 1980, p. 110; M.A. Gusakov (ed.), *Organizatsionnye formy sviazi nauki s proizvodstvom v promyshlennosti*, Nauka, Leningrad, 1980, p. 81; *Metodologicheskie problemy nauchno-tekhnicheskogo progressa. Voprosy integratsii nauki i proizvodstva*, Nauka, Novosibirsk, 1987, p. 145.
41 Bakunin, *Deiatel'nost' KPSS*, p. 176.
42 Amann and Cooper, *Industrial Innovation*, pp. 192–3.

43 *Literaturnaia gazeta* 26 March 1975, p. 10.
44 *Pravda* 26 September 1971, p. 3; *Literaturnaia gazeta* 28 June 1976, p. 10.
45 Amann and Cooper, *Industrial Innovation*, p. 192.
46 *Pravda* 12 June 1985, p. 2.
47 *Ekonomicheskaia gazeta* no.11 (1986), p. 26.

Chapter 6 The research institute

1 *Vestnik statistiki* no.7 (1988), p. 64.
2 *VAN* no.10 (1987), p. 38.
3 *Pravda* 15 October 1987, p. 3.
4 *Pravda* 9 April 1987, p. 3; *Ekonomicheskaia gazeta* no.15 (1987), p. 23. For a geographical distribution of scientific output based on publication data, see *VAN* no.5 (1988), p. 73.
5 *Naukovedenie i informatika* no.21 (1980), p. 24.
6 V.I. Duzhenkov, *Problemy organizatsii nauki (Regional'nye aspekty)*, Nauka, Moscow, 1978, p. 81.
7 Duzhenkov, *Problemy organizatsii*, p. 175.
8 *Pravda* 7 July 1970, p. 3; *Izvestiia* 29 October 1987, p. 3.
9 *Izvestiia* 25 August 1986, p. 2.
10 *VAN* no.4 (1984), p. 63.
11 *VAN* no.3 (1976), p. 6; *Ekonomicheskaia gazeta* no.24 (1987), p. 5.
12 *Literaturnaia gazeta* 26 February 1966, pp. 1–2.
13 Duzhenkov, *Problemy organizatsii*, p. 175.
14 In the Siberian Academy in the late 1970s only 6.5 per cent of deputy directors had no advanced degree, 44.6 per cent were candidates of science, and 48.9 per cent doctors of science. A.K. Romanov, L.A. Androsova, and A.F. Felinger, *Nauchnye kadry Sibirskogo otdeleniia AN SSSR. Metody i rezul'taty statisticheskogo issledovaniia*, Nauka, Novosibirsk, 1979, p. 48.
15 V. Dorofeeva and V. Dorofeev, *Vremia, uchenye, sversheniia*, Politizdat, Moscow, 1975, p. 281.
16 *Leningradskaia pravda* 7 May 1986, p. 2.
17 N. Kaplan, 'Research administration and the administrator: USSR and US', *Administrative Science Quarterly* vol.6 (1961), pp. 56–7.
18 *VAN* no.11 (1975), p. 28.
19 *VAN* no.11 (1975), p. 28.
20 *VAN* no.11 (1975), p. 29.
21 A.I. Shcherbakov, *Effektivnost' nauchnoi deiatel'nosti v SSSR*, Ekonomika, Moscow, 1982, p. 219.
22 Shcherbakov, *Effektivnost'*, p. 217.
23 *Organizatsiia promyshlennykh issledovanii i razrabotok novoi tekhniki*, Lenizdat, Leningrad, 1978, p. 170.
24 Romanov, *Nauchnye kadry*, p. 48.
25 See, for example, A. Zakharov, scientific secretary of the Institute of Space Research, being described as a plasma physicist in *Science in the USSR* no.4 (1986), p. 4. See also the work on water resources of the scientific secretary of the Institute of Geography of the USSR Academy, A.V. Beliaev, as described in *Priroda* no.3 (1987), p. 60.

26 *I.P. Bardin v vospominaniiakh sovremennikov*, Nauka, Moscow, 1985, p. 182.
27 R. Berg, *Sukhovei. Vospominaniia genetika*, Chalidze, New York, 1983, pp. 237–8.
28 M.Ya. Azbel, *Refusenik. Trapped in the Soviet Union*, Hamish Hamilton, London, 1982, p. 279.
29 *Izvestiia* 7 June 1961, p. 4.
30 Berg, *Sukhovei*, p. 309; 'The case of Igor Melchuk', *Survey* vol.23, no.2 (103) (Spring 1977–8), p. 127; *Kommunist (Litva)* no.2 (1970), p.58; *Kommunist* no.6 (1972), p.94.
31 *Pravda* 3 August 1986, p. 3.
32 *Partiinaia zhizn'* no.21 (1976), p. 72.
33 *Izvestiia* 26 August 1969, p. 3.
34 See the cases of Mel'chuk and Berg. 'The case of Igor Melchuk', p. 127; Berg, *Sukhovei*, p. 309.
35 See the conflict of views between the president of the Academy Marchuk and a group of academy social scientists. *Izvestiia* 22 March 1987, p. 2; *Literaturnaia gazeta* 8 July 1987, p. 12.
36 *VAN* no.12 (1987), p. 24.
37 *VAN* no.12 (1987), pp. 42–3.
38 E. Belitsky, *The Central Institute of Mathematical Economics*, Soviet Institutions Series, Paper No.1, The Hebrew University of Jerusalem, 1974, p. 67; *Literaturnaia gazeta* 24 March 1976, p. 11.
39 Belitsky, *Central Institute*, p. 67.
40 *VAN* no.11 (1975), p. 25.
41 *VAN* no.7 (1961), p. 41; *Izvestiia* 25 January 1962, p. 3.
42 *Leningradskaia pravda* 16 August 1986, p. 2.
43 *Ekonomicheskaia gazeta* no.47 (November 1979).
44 Iu.M. Kanygin and V.A. Botvin, *Problemy razvitiia i ispol'zovaniia nauchnogo potentsiala krupnykh gorodov*, Naukova dumka, Kiev, 1980, p. 84.
45 V.A. Trapeznikov (ed.), *Upravlenie razvitiem nauki i tekhniki*, Ekonomika, Moscow, 1980, p. 53; M.A. Gusakov (ed.), *Organizatsionnye formy sviazi nauki s proizvodstvom v promyshlennosti*, Nauka, Leningrad, 1980, pp. 109–10.
46 *Vestnik statistiki* no.7 (1988), p. 67.
47 See Marchuk's comments in *Izvestiia* 22 March 1987, p. 2.
48 *VAN* no.6 (1980), p. 21.
49 *Naukovedenie i informatika* no.21 (1980), p. 29.
50 V.A. Pokrovskii, *Uskorenie nauchno-tekhnicheskogo progressa. Organizatsiia i metody*, Ekonomika, Moscow, 1983, p. 80.
51 *Izvestiia* 30 November 1985, p. 3.
52 Kanygin and Botvin, *Problemy razvitiia*, p. 152; *Izvestiia* 21 June 1986, p. 2; 25 August 1987, p. 2.
53 S. Fortescue, 'The Soviet Academy of Sciences under Gorbachev', *Australian Slavonic and East European Studies* vol.1, no.2 (1987), p. 21.
54 V.I. Kudashov, E.S. Kukolko, and P.D. Gorodets, *Upravlenie issledovaniiami i razrabotkami v akademicheskom nauchno-tekhnicheskom*

komplekse, Nauka i tekhnika, Minsk, 1986, p. 105; V.I. Nedil'ko, *Intensifikatsiia nauki: soderzhanie i planovo-finansovye faktory*, Nauka i tekhnika, Minsk, 1984, pp. 74, 85.

55 *VAN* no.11 (1975), p. 25.

56 Shcherbakov, *Effektivnost'*, p. 98.

57 *Organizatsiia i razvitie otraslevykh nauchno-issledovatel'skikh institutov Leningrada, 1917–1977*, Nauka, Leningrad, 1979, p. 209.

58 I. Zemtsov, *IKSI. The Moscow Institute of Applied Social Research. A note on the development of sociology in the USSR*, Soviet Institutions Series, Paper No.6 The Hebrew University of Jerusalem, April 1976, p. 45.

59 Berg, *Sukhovei*, pp. 259–61.

60 'Subsidiary farms' are farms connected to enterprises and worked by enterprise staff. 'Food orders' consist of food parcels – often a tantalising mixture of deficit and surplus products – obtained by work collectives from food wholesalers and sold on to their staff.

61 *Izvestiia* 14 June 1988, p. 1.

62 S. Fortescue, *The Communist Party and Soviet Science*, Macmillan, London and Basingstoke, 1986, chapter 4; 'Research party organisations and the right of control', *Soviet Studies* vol.35, no.2 (April 1983), pp. 175–95.

63 S. Fortescue, 'Party membership in Soviet research institutes', *Soviet Union/Union Sovietique* vol.11, pt 2 (1984), pp. 144–7.

64 *Kadry – reshaiushchaia sila partiinogo rukovodstva*, Mysl', Moscow, 1984, p. 136; *Partiia ukrepliaet svoi riady* (Mysl', Moscow, 1985), p. 81.

65 R.P. Suttmeier, 'Socialist science policy', *Problems of Communism* vol.26, no.4 (July–August 1977), p. 75.

66 S. Fortescue, 'Party secretaries in Soviet research institutes: divided loyalties?', *Politics* (Australia) vol. 18, no.1 (May 1983), pp. 73–83.

67 S. Fortescue, 'Project planning in Soviet R&D', *Research Policy* vol.14 (1985), pp. 267–82.

68 R. Amann and J. Cooper (eds), *Industrial Innovation in the Soviet Union*, Yale University Press, New Haven, CN, and London, 1982, p. 464.

69 *VAN* no.7 (1984), p. 81; no.7 (1985), p. 18; *Leningradskaia pravda* 22 February 1986, p. 1.

70 *Izvestiia* 13 January 1986, p. 2.

71 *Izvestiia* 13 January 1986, p. 2; *Metodologicheskie problemy razvitiia ekonomicheskikh sviazei v sisteme 'Nauka – proizvodstvo'*, Moscow, 1983, p. 17.

72 *VAN* no.1 (1987), p. 36.

73 *Izvestiia* 22 March 1987, p. 2.

74 *Leningradskaia pravda* 16 September 1986, p. 2. See also the comments of Velikhov in *VAN* no.12 (1986), pp. 17–18.

75 *Izvestiia* 13 January 1986, p. 2.

76 *EKO* no.7 (1985), p. 86.

77 *Metodologicheskie problemy razvitiia*, p. 22.

78 *Leningradskaia pravda* 30 October 1985, p. 2.

79 *Literaturnaia gazeta* 18 March 1987, p. 11; *Izvestiia* 3 May 1987, p. 2; *Ekonomicheskaia gazeta* no.12 (1987), pp. 14–15.

80 *Izvestiia* 15 May 1988, p. 2.

81 For details of 'Otklik' in Kiev, working on electromagnetic medical diagnostics, see *Pravda* 7 July 1987, p. 3; *Izvestiia* 3 March 1988, p. 3.
82 *Izvestiia* 24 May 1988.
83 *Partiinaia zhizn'* no.20 (1987), p. 45. For a history of 'Fakel', see J. Lowenhardt, '"The tale of the torch". Scientist entrepreneurs in the Soviet Union', *Survey* vol.20, no.4 (93) (Autumn 1974), pp. 113–21.
84 *Izvestiia* 21 March 1985, p. 2; 8 March 1987, p. 2; 15 March 1987, p. 2.
85 *Izvestiia* 31 May 1988, p. 2.
86 *Moscow News* no.30 (1987), p. 11. For some rather complex detail on the pricing and supply procedures of a Moldavian cooperative, see *Izvestiia* 31 January 1987, p. 2.
87 See R.A. Lewis, 'Hierarchy and technological innovation in Soviet industry: the science–production associations', *Minerva* vol. 22, no.2 (Summer 1984), pp. 129–59. For a comprehensive Soviet source, see K.I. Taksir, *Nauchno-proizvodstvennye ob"edineniia*, Nauka, Moscow, 1977.
88 *Resheniia partii i pravitel'stva po khoziaistvennym voprosam. Sbornik dokumentov. Tom 11, Noiabr' 1975 g. – iun' 1977 g.*, Politizdat, Moscow, 1977, pp. 65–113.
89 See Ryzhkov at the 27th Party Congress and the Thirteenth Five-Year Plan.
90 *Pravda* 24 October 1986, p. 1; *Izvestiia* 22 February 1988, p. 1.
91 Pokrovskii, *Uskorenie NTP*, p. 65; *Voprosy ekonomiki* no.9 (1979), p. 128.
92 L.E. Nolting, *The 1968 Reform of Scientific Research, Development and Innovation in the USSR*, US Department of Commerce, Washington, DC, 1976, p. 16.
93 *Sotsialisticheskaia industriia* 22 February 1986, p. 2; *Izvestiia* 25 May 1986, p. 2.
94 *Pravda* 24 October 1986, p. 1.
95 *Pravda* 11 October, 1984.
96 *Ekonomicheskaia gazeta* no.49 (1986), p. 2.
97 *Ekonomicheskaia gazeta* no.3 (1987), p. 10; *Pravda* 24 October 1986, p. 1.
98 *VAN* no.2 (1987), p. 71.
99 *Sotsialisticheskaia industriia* 27 March 1988, p. 2; 5 April 1988, p. 2.
100 *Izvestiia* 19 August 1986, p. 2; *Pravda* 12 March 1986, p. 2.
101 *Pravda* 22 July 1987, p. 2.
102 *Sotsialisticheskaia industriia* 20 December 1985, p. 2.
103 S. Fortescue, *The Academy Reorganized. The R&D role of the Soviet Academy of Sciences since 1961*, Occasional Paper No.17, Department of Political Science, Research School of Social Sciences, Australian National University, Canberra, 1983, pp. 52–4.
104 *Pravda* 13 December 1985, p. 1.
105 *Izvestiia* 23 January 1985, p. 1.
106 *Sotsialisticheskaia industriia* 14 February 1987, p. 2; *Ekonomicheskaia gazeta* no.17 (1987), p. 10.
107 *Ekonomicheskaia gazeta* no.40 (1987), p. 14.
108 *Ekonomicheskaia gazeta* no.34 (1986), p. 7; *Pravda* 8 February 1987, p. 4.

109 *Leningradskaia pravda* 14 February 1986, p. 2.
110 *Voprosy izobretatel'stva* no.9 (1986), p. 11; *Sotsialisticheskaia industriia* 20 December 1985, p. 2; 14 February 1987, p. 2; *Ekonomicheskaia gazeta* no.17 (1987), p. 10.
111 *Ekonomicheskaia gazeta* no.3 (1986), p. 9.
112 *Voprosy izobretatel'stva* no.9 (1986), p. 11.
113 *Izvestiia* 17 April 1986, p. 2.
114 *Voprosy izobretatel'stva* no.9 (1986), p. 11.
115 *Sotsialisticheskaia industriia* 1 April 1986, p. 2; 2 September 1986, p. 2; *Ekonomicheskaia gazeta* no.34 (1986), p.7.
116 *Sotsialisticheskaia industriia* 31 July 1987, p. 2.
117 *Izvestiia* 10 July 1987, p. 2.
118 *Sotsialisticheskaia industriia* 31 July 1987, p. 2.
119 *Sotsialisticheskaia industriia* 31 July 1987, p. 2.
120 *Ekonomicheskaia gazeta* no.40 (1987), p. 14.
121 *Izvestiia* 19 August 1986, p. 2.
122 *Ekonomicheskaia gazeta* no.17 (1987), p. 10. See also *Pravda* 15 November 1986, p. 2.
123 *Pravda* 24 March 1987, p. 1.
124 *Sotsialisticheskaia industriia* 14 February 1987, p. 2.
125 *VAN* no.7 (1986), p. 8.
126 *Izvestiia* 3 October 1986, p. 2; *VAN* no.2 (1987), p. 40.
127 *Sotsialisticheskaia industriia* 14 February 1987, p. 2.
128 Fortescue, 'The Soviet Academy of Sciences', pp. 19–22. See Alferov's comments in *Izvestiia* 17 November 1987, p. 2.
129 *Ekonomicheskaia gazeta* no.17 (1987), p. 10. My information shows that 'Antikor' was originally headed by a GKNT, not Academy institute.
130 *Pravda* 26 February 1988, p. 2.
131 *VAN* no.12 (1987), p. 19.
132 *Izvestiia* 20 October 1987, p. 5; *VAN* no.1 (1988), p. 19.
133 *Ekonomicheskaia gazeta* no.40 (1987), p. 14.
134 *Izvestiia* 9 August 1988, p. 2; *Ekonomicheskaia gazeta* no.15 (1988), p. 7; *Moscow News* 23 October 1988, p. 7.
135 *Pravda* 20 May 1988, p. 2.

Chapter 7 The planning of Soviet science

1 *Sotsialisticheskaia industriia* 10 February 1987, p. 2.
2 L.E. Nolting, *The Planning of Research, Development and Innovation in the USSR*, Foreign Economic Report No.14, US Department of Commerce, Washington, DC, 1978, pp. 7–10.
3 M. Kuchment and H. Balzer, *A Net Assessment of Soviet R&D Capabilities: information and insights from the third emigration*, Russian Research Center, Harvard, 1981, p. 22.
4 B.V. Shcherbitskii and V.S. Tarasovich (eds), *Planovoe upravlenie razvitiem nauki i tekhniki v soiuznoi respublike*, Naukova dumka, Kiev, 1981, p. 138.
5 *Izvestiia* 3 July 1984, p. 2. See also *Osnovy upravleniia naukoi. Ekonomicheskie problemy*, Nauka, Moscow, 1983, p. 57.

6 Iu.I. Berliner, *Vnedrenie nauchno-tekhnicheskikh razrabotok. Problemy planirovaniia*, Ekonomika, Moscow, 1985, p. 42. See also *Pravda* 7 July 1970, p. 3.
7 *Voprosy teorii i praktiki upravleniia i organizatsiia nauki*, Nauka, Moscow, 1975, p. 74; *Literaturnaia gazeta* 17 October 1979, p. 13.
8 A later request was for a research reactor to be built in Tbilisi. This request was met, and Kurchatov throughout the period of its design and construction required regular reports on progress from Andronikashvili and used his power to break down bottlenecks and delays. E. Andronikashvili, *Nachinaiu s El'brusa (Tvorcheskie portrety uchenykh)*, Metsnieraba, Tbilisi, 1982, pp. 22–4, 58–62, 77–8.
9 Andronikashvili, *Nachinaiu*, pp. 236–9.
10 I. Golovin, *Igor Kurchatov*, Mir, Moscow, 1969, p. 112.
11 R.J. Marsh, *Soviet Fiction since Stalin. Science, politics and literature*, Croom Helm, London and Sydney, 1986, p. 291.
12 *Vospominaniia ob akademike L.A. Artsimoviche*, Nauka, Moscow, 1981, pp. 32–3, 40–1.
13 *Sotsialisticheskaia industriia* 23 February 1985, p. 4; 10 February 1987, p. 2.
14 *Teoriia i praktika organizatsii kompleksnykh issledovanii*, Nauka, Leningrad, 1986, p. 49.
15 S.R. Mikulinskii and R. Rikhta (eds), *Sotsializm i nauka*, Nauka, Moscow, 1981, p. 228.
16 For example, see Academy vice-president Kotel'nikov in *VAN* no.11 (1986), p. 32. See also *Izvestiia* 11 November 1987, p. 2; *Ekonomicheskaia gazeta* no.8 (1988), p. 16.
17 Academician Sagdeev has stated to a Western audience that the intention is to have 20–30 per cent of the science budget allocated through grants. R. Sagdeev, 'Perestroika and the scientific intelligentsia', Meeting report, Kennan Institute for Advanced Russian Studies, Washington, DC, 16 November 1988.
18 L.L. Veger, *Ekonomicheskii effekt i upravlenie NIOKR*, Nauka, Moscow 1985, p. 94.
19 *Voprosy teorii i praktiki*, p. 67.
20 *VAN* no.11 (1986), p. 31.
21 Berliner, *Vnedrenie*, p. 60.
22 *Ekonomicheskaia gazeta* no.14 (1988), p. 23.
23 *Pravda* 3 August 1966, p. 2; 14 August 1969, p. 2; *VAN* no.3 (1976), p. 10.
24 *Finansy SSSR* no.10 (1981), p. 36.
25 *Trud* 4 January 1973, p. 2; *Sotsialisticheskaia industriia* 10 May 1987, p. 3. For other examples, see *Literaturnaia gazeta* no.36 (1981), p. 12; no.46 (1982), p. 8.
26 *Sovetskaia Rossiia* 12 March 1986, p. 3.
27 *VAN* no.3 (1957), p. 7.
28 *VAN* no.6 (1978), p. 64.
29 L.E. Nolting, *Financing of Research, Development and Innovation in the USSR*, Foreign Economic Report No. 22, Department of Commerce, Washington, DC, September 1985, pp. 31–2. For a good summary of the

system as introduced into the pioneering electrotechnical industry, see L.E. Nolting, *Management and Financing of Research, Development, and Innovation in the Soviet Electrotechnical Industry*, Center for International Research Staff Paper No.43, US Bureau of the Census, Washington, DC, August 1988.

30 V.A. Pokrovskii, *Uskorenie nauchno-tekhnicheskogo progressa. Organizatsiia i metody*, Ekonomika, Moscow, 1983, p. 69.

31 *Finansy SSSR* no.10 (1981), p. 35.

32 Nolting, *Financing of RDI*, p. 17; V.P. Rassokhin, *Mekhanizm vnedreniia dostizhenii nauki. Politika, upravlenie, pravo*, Nauka, Moscow, 1985, pp. 105, 220.

33 For details on the new approach, see S. Fortescue, 'Project planning in Soviet R&D', *Research Policy* vol.14 (1985), pp. 267–82.

34 *VAN* no.6 (1978), p. 63; Pokrovskii, *Uskorenie NTP*, p. 118; *Sotsialisticheskaia industriia* 13 June 1985, p. 1.

35 *VAN* no.8 (1984), p. 37.

36 *Voprosy ekonomiki* no.2 (1986), p. 31.

37 Berliner, *Vnedrenie*, p. 22.

38 *Zaria Vostoka* 5 June 1982, pp. 1, 3; I.I. Sigov (ed.), *Kompleksnaia programma nauchno-tekhnicheskogo progressa v regione*, Nauka, Leningrad, 1983, p. 90.

39 *Sotsialisticheskaia industriia* 8 January 1988, p. 2; 27 January 1988, p. 1.

40 *Sobranie postanovlenii SSSR* no.48 (1987), art.158, para.4.

41 *Sobranie postanovlenii SSSR* no.48 (1987), art.158, para.10.

42 *Ekonomicheskaia gazeta* no.10 (1988), p. 14.

43 *Ekonomicheskaia gazeta* no.2 (1987), pp. 6–7.

44 *Sotsialisticheskaia industriia* 26 February 1988, p. 2.

45 *Sobranie postanovlenii SSSR* no.48 (1987), art.158, para.6.

46 *Sobranie postanovlenii SSSR* no.48 (1987), art.158, para.8; *Sotsialisticheskaia industriia* 26 February 1988, p. 2.

47 *Ekonomicheskaia gazeta* no.22 (1987), p. 9.

48 *Pravda* 8 February 1988, p. 3; 3 March 1988, p. 2; 31 May 1988, p. 2; *Ekonomicheskaia gazeta* no.6 (1988), p. 6.

49 *Izvestiia* 29 December 1987, p. 2; *Ekonomicheskaia gazeta* no.6 (1988), p.6.

50 *Pravda* 8 February 1988, p. 3; 3 March 1988, p. 2.

51 For a good account of the traditional lack of working capital in Soviet institutes, see Nolting, *Management*, pp. 65–7.

52 *VAN* no.5 (1988), pp. 20–1.

53 *Ekonomicheskaia gazeta* no.17 (1986), p. 6.

54 *Izvestiia* 18 October 1987, p. 4; *VAN* no.5 (1988), p. 9.

55 *Ekonomicheskaia gazeta* no.2 (1988), p. 8; *Pravda* 3 March 1988, p. 2.

56 *Ekonomicheskaia gazeta* no.6 (1988), p. 6.

57 *Ekonomicheskaia gazeta* no.6 (1988), p. 6.

58 *Pravda* 3 March 1988, p. 2. For details on another institute which seems to have made the transition successfully, the Groznyi Oil Research Institute, see *Ekonomicheskaia gazeta* no.15 (1988), p. 6.

59 *Izvestiia* 14 June 1988, p. 1; 9 August 1988, p. 2.

60 *Sobranie postanovlenii SSSR* no.48 (1987), art.158, paras 15 and 17.

61 *Ekonomicheskaia gazeta* no.2 (1988), p. 9; no.6 (1988), p. 6.
62 *Izvestiia* 2 July 1987, p. 2; *VAN* no.5 (1988), pp. 7–8.
63 *Sotsialisticheskaia industriia* 26 February 1988, p. 2.
64 *Pravda* 17 April 1988, p. 3; *Kommunist* no.17 (1987), p. 61; *Ekonomicheskaia gazeta* no.6 (1988), p. 6.
65 *Ekonomicheskaia gazeta* no.15 (1988), p. 6.

Chapter 8 Research managers

1 *VAN* no.7 (1961), p. 41; *Izvestiia* 25 January 1962, p. 3.
2 *Izvestiia* 15 May 1988, p. 2.
3 L.L. Lubrano and S.G. Solomon (eds), *The Social Context of Soviet Science*, Westview, Boulder, CO, and Dawson, Folkestone, 1980, pp. 39–40.
4 S.R. Mikulinskii and R. Rikhta (eds), *Sotsializm i nauka*, Nauka, Moscow, 1981, pp. 230–1.
5 *VAN* no.12 (1961), p. 41.
6 A.I. Sliusarenko, *KPSS i progress nauki v razvitom sotsialisticheskom obshchestve*, Vishcha shkola, Kiev, 1978, p. 82.
7 A.K. Romanov, L.A. Androsova, and A.F. Felinger, *Nauchnye kadry Sibirskogo otdeleniia AN SSSR. Metody i rezul'taty statisticheskogo issledovaniia*, Nauka, Novosibirsk, 1979, p. 48.
8 *Izvestiia* 24 August 1987, p. 2.
9 *Pravda* 17 August 1987, p. 2.
10 *Pravda* 12 November 1972, p. 2.
11 *Izvestiia* 16 January 1986, p. 6.
12 *Voprosy filosofii* no.3 (1970), p. 81; *Pravda* 12 November 1972, p. 2.
13 *Izvestiia* 21 July 1988, p. 3; 14 September 1988, p. 3.
14 P.G. Belkin, E.N. Emel'ianov, and M.A. Ivanov, *Sotsial'naia psikhologiia nauchnogo kollektiva*, Nauka, Moscow, 1987, p. 115.
15 V.V. Kosolapov and A.N. Shcherban', *Optimizatsiia nauchno-issledovatel'skoi deiatel'nosti*, Naukova dumka, Kiev, 1971, pp. 133–4; M. Popovsky, *Upravliaemaia nauka*, Overseas Publications Interchange, London, 1978, pp. 51–2.
16 *Moscow News* no.30 (1987), p. 12; *Sotsialisticheskaia industriia* 31 May 1987, p. 3.
17 *Sovetskaia Rossiia* 4 November 1984, p. 3.
18 *Izvestiia* 7 April 1987, p. 3.
19 A.A. Zvorykin (ed.), *Nauchnyi kollektiv. Opyt sotsiologicheskogo issledovaniia*, Nauka, Moscow, 1980, p. 103.
20 Zvorykin, *Nauchnyi kollektiv*, p. 104. For another survey reporting very different findings, although without details, see *Sotsialisticheskaia industriia* 10 February 1987, p. 2.
21 G.A. Lakhtin, *Upravlenie v nauchnom uchrezhdenii*, Energoatomizdat, Moscow, 1983, p. 113.
22 *Izvestiia* 20 April 1987, p. 2.
23 Popovsky, *Upravliaemaia nauka*, p. 49.
24 *Novyi mir* no.4 (1988), p. 181.
25 *Moscow News* no.33 (1987), p. 10.
26 For more details on the plate tectonics controversy, see S. Fortescue, *The*

Communist Party and Soviet Science, Macmillan, London and Basingstoke, 1986, pp. 22–4.
27 *Physics Today* January 1984, p. 9.
28 *Literaturnaia gazeta* 11 March 1987, p. 11.
29 *Partiinaia zhizn'* no.21 (1976), p. 72; *Literaturnaia gazeta* 11 March 1987, p. 11.
30 *Izvestiia* 18 August 1985, p. 3. See also *Izvestiia* 7 April 1987, p. 3.
31 *Sovetskaia Rossiia* 20 August 1985, p. 2; 16 May 1986, p. 3.
32 For example, *Sotsialisticheskaia industriia* 4 May 1988, p. 2; 28 August 1988, p. 1; *Pravda* 11 October 1985, p. 2.
33 For examples, see *Pravda* 24 May 1987, p. 1; *Sotsialisticheskaia industriia* 2 September 1987, p. 2; *Moscow News* no.30 (1987), p. 12. For more general comments on the problems involved with the attestation campaign, including the squabbles, gossip, and denunciations, see *VAN* no.1 (1987), p. 67; *Sotsialisticheskaia industriia* 15 May 1987, p. 2; *Ekonomicheskaia gazeta* no.24 (1987), p. 5.
34 *Izvestiia* 7 April 1987, p. 3.
35 *Sotsialisticheskaia industriia* 20 December 1987, pp. 2–3. For other cases in which criminal charges have been brought against a director's victim, see *Sotsialisticheskaia industriia* 19 March 1988, p. 2; *Literaturnaia gazeta* 23 March 1988, p. 11.
36 *Literaturnaia gazeta* 23 March 1988, p. 11. See also *Moscow News* no.30 (1987), p. 12.
37 *Izvestiia* 30 May 1987, p. 3; 18 October 1987, p. 6.
38 R. Berg, *Sukhovei. Vospominaniia genetika*, Chalidze, New York, 1983, pp. 227–32.
39 For details, see Fortescue, *The Communist Party*, pp. 44–5.
40 B.Ia. Frenkel', *Iakov Il'ich Frenkel'*, Nauka, Moscow–Leningrad, 1966, p. 213.
41 M.Ya. Azbel, *Refusenik. Trapped in the Soviet Union*, Hamish Hamilton, London, 1982, p. 72.
42 *Physics Today* May 1987, pp. 79–80.
43 *Priroda* no.5 (1987), p. 85.
44 *Vospominaniia o I.E. Tamme*, 2nd edn, Nauka, Moscow, 1986, p. 248.
45 Berg, *Sukhovei*, pp. 226–71, 320–2; Fortescue, *The Communist Party*, pp. 48–9, 192.
46 *Sovetskoe gosudarstvo i pravo* no.1 (1987), p. 27; *Izvestiia* 10 May 1988, p. 2.
47 *Pravda* 8 March 1967, p. 3.
48 L.L. Lubrano, 'Soviet science specialists: professional roles and policy involvement', in R.B. Remnek (ed.), *Social Scientists and Policy Making in the USSR*, Praeger, New York, 1977, pp. 67–8.
49 *Akademik I.I. Artobolevskii*, Znanie, Moscow, 1983, p. 75.
50 *Ideino-politicheskoe vospitanie nauchno-tekhnicheskogo intelligentsii. Opyt i problemy*, Nauka, Moscow, 1982, p. 47; *Zaria Vostoka* 30 March 1979, p. 2.
51 Lakhtin, *Upravlenie*, p. 114.
52 *Partiinaia zhizn'* no.21 (1987), p. 30.

53 *Ekonomicheskaia gazeta* no.6 (1988), p. 7.
54 *Izvestiia* 18 November 1987, p. 3.
55 *VAN* no.3 (1976), p. 6; *Sotsialisticheskaia industriia* 3 December 1986, p. 1; *Ekonomicheskaia gazeta* no.24 (1987), p. 5.
56 *Izvestiia* 22 March 1987, p. 2.
57 *Zaria Vostoka* 3 November 1981, p. 3.
58 *Izvestiia* 1 December 1985, p. 3.
59 *Sobranie postanovlenii SSSR* no.7 (1988), art.15; *VAN* no.7 (1987), p. 16.
60 S. Fortescue, 'The Soviet Academy of Sciences under Gorbachev', *Australian Slavonic and East European Studies* vol. 1, no.2 (1987), pp. 11–14.
61 *Izvestiia* 12 June 1988, p. 2.
62 *Izvestiia* 22 March 1987, p. 2.
63 *Sotsialisticheskaia industriia* 20 December 1987, p. 2.
64 *Literaturnaia gazeta* 8 July 1987, p. 12.
65 *Russkaia Mysl'* 11 March 1988. For a case of the Department apparently approving the popular choice, see *VAN* no.3 (1988), p. 79.
66 *VAN* no.2 (1988), p. 4.
67 *Izvestiia* 22 March 1987, p. 2.

Chapter 9 The staff

1 *Sotsialisticheskaia industriia* 10 February 1987, p. 2.
2 S. Fortescue, *The Communist Party and Soviet Science*, Macmillan, London and Basingstoke, 1986, pp. 35–6.
3 Fortescue, *The Communist Party*, p. 36; *Sovetskaia Rossiia* 29 March 1985, p. 3; *Leningradskaia pravda* 8 July 1986, p. 2; *Sotsialisticheskaia industriia* 20 December 1987, p. 2. It should be mentioned that not all sources agree that such a problem exists. See *VAN* no.12 (1985), p. 12; *Uchenyi i nauchnyi kollektiv: sotsial'nye aspekty deiatel'nosti*, Progress, Moscow, 1986, p. 138.
4 *Sotsialisticheskaia industriia* 10 February 1987, p. 2.
5 A.I. Shcherbakov, *Effektivnost' nauchnoi deiatel'nosti v SSSR*, Ekonomika, Moscow, 1982, p. 148. For another survey with similar results, see *Voprosy teorii i praktiki upravleniia i organizatsiia nauki*, Nauka, Moscow, 1975, p. 248.
6 P.G. Belkin, E.N. Emel'ianov, and M.A. Ivanov, *Sotsial'naia psikhologiia nauchnogo kollektiva*, Nauka, Moscow, 1987, p. 114.
7 Belkin, *Sotsial'naia psikhologiia*, p. 151.
8 *VAN* no. 1 (1988), p. 44.
9 *Voprosy teorii i praktiki*, p. 248.
10 *Teoriia i praktika organizatsii kompleksnykh issledovanii*, Nauka, Leningrad, 1986, p. 150.
11 V.V. Kosolapov and A.N. Shcherban', *Optimizatsiia nauchno-issledovatel'skoi deiatel'nosti*, Naukova dumka, Kiev, 1971, p. 139.
12 For a good example, see *Pravda* 5 January 1987, p. 3.
13 *Problemy rukovodstva nauchnym kollektivom. Opyt sotsial'no-psikhologicheskogo issledovaniia*, Nauka, Moscow, 1982, p. 91.
14 Shcherbakov, *Effektivnost'*, p. 148.
15 *Vosprosy teorii i praktiki*, pp. 248–9.
16 Kosolapov, *Optimizatsiia*, p. 138.

17 Fortescue, *The Communist Party*, pp. 35–7.
18 Fortescue, *The Communist Party*, p. 35.
19 See S. Fortescue, 'Recent developments in the administration of branch military research', in C. Sinclair (ed.), *The Status of Soviet Civil Science*, Nijhoff, Dordrecht, 1987, p. 111.
20 See scientists complaining about their treatment in the press in *Pravda* 19 July 1986, p. 3; *Literaturnaia gazeta* 11 March 1987, p. 11.
21 For example, *Kommunist* no.11 (1985), pp. 63–73; *Izvestiia* 21 June 1988, p. 2; R. Sagdeev, 'Perestroika and the scientific intelligentsia', Meeting report, Kennan Institute for Advanced Russian Studies, Washington, DC, 16 November 1988.
22 See Fortescue, *The Communist Party*, pp. 36–7. For a recent detailed examination of the treatment of science in Soviet literature, see R.J. Marsh, *Soviet Fiction since Stalin. Science, politics and literature*, Croom Helm, London and Sydney, 1986.
23 Belkin, *Sotsial'naia psikhologiia*, pp. 151–2.
24 I report a number of such surveys in *The Communist Party*, pp. 46–7. See also I.D. Andreev, *Puti povysheniia effektivnosti nauchnogo truda*, Nauka, Moscow, 1985, p. 105; *Nauchno-tekhnicheskaia intelligentsiia razvitogo sotsializma*, Kemerovskii gos. un-t, Kemerovo, 1985, p. 51; *Nature* 26 February 1987, p. 467.
25 With some even suggesting that such 'individualism' contributes to dissident tendencies. See the Academy's chief ideologist Petr Fedoseev as quoted in Fortescue, *The Communist Party*, p. 49.
26 Belkin, *Sotsial'naia psikhologiia*, p. 152.
27 A.A. Zvorykin (ed.), *Nauchnyi kollektiv. Opyt sotsiologicheskogo issledovaniia*, Nauka, Moscow, 1980, p. 55; *Sotsialisticheskaia industriia* 23 February 1985, p. 4.
28 Belkin, *Sotsial'naia psikhologiia*, pp. 147–9, 159.
29 The branch teaching institutes have to be distinguished from branch research institutes. The latter are usually described as *'nauchno-issledovatel'skii institut'* (scientific research institute). The former are simply called '*institut*', for example, the Institute of Steel and Alloys.
30 M.J. Berry (ed.), *Science and Technology in the USSR*, Longman, Harlow, 1988, p. 58.
31 V.G. Afanas'ev, *Nauchno-tekhnicheskaia revoliutsiia, upravlenie, obrazovanie*, Politizdat, Moscow, 1972, p. 377.
32 *Voprosy teorii i praktiki*, p. 282; *Voprosy istorii KPSS* no.11 (1986), p. 15.
33 *Sovetskaia Rossiia* 28 September 1985, p. 1.
34 *Kommunist* no.18 (1968), p. 43.
35 Belkin, *Sotsial'naia psikhologiia*, p. 152.
36 L.V. Levshin, *Aleksandr Nikolaevich Terenin 1896–1967*, Nauka, Moscow, 1985, p. 41; A.P. Grinberg and B.Ia. Frenkel', *Igor' Vasil'evich Kurchatov v Fiziko-Tekhnicheskom Institute*, Nauka, Leningrad, 1984, p. 114.
37 *Sotsialisticheskaia industriia* 20 December 1987, p. 2.
38 *VAN* no.7 (1985), p. 13.
39 *VAN* no.8 (1979), p. 75.
40 *VAN* no.8 (1987), p. 44.

41 *Voprosy teorii i praktiki*, pp. 177–81.
42 *Problemy nauchnogo kommunizma* no.7 (1973), pp. 139–40.
43 *Izvestiia* 23 July 1986, p. 2. For more data, see *Kommunist* no.17 (1987), p. 55; *Sotsialisticheskaia industriia* 4 June 1986, p. 2.
44 *Kommunist* no.7 (1986), p. 38; *VAN* no.7 (1987), p. 61. On poor housing and heating and problems with telephones in Troitsk, also outside Moscow, see *Izvestiia* 29 January 1986, p. 6.
45 *Izvestiia* 22 March 1987, p. 2.
46 *Pravda* 18 December 1986, p. 2.
47 *Pravda* 20 August 1987, p. 2.
48 *VAN* no.3 (1984), p. 15.
49 Listed in *VAN* no.12 (1987), pp. 124–5.
50 See the saga of the Urals Science Centre, in *Pravda* 22 September 1974, p. 2; *Izvestiia* 18 February 1984, p. 2; *Pravda* 24 April 1986, p. 2; *VAN* no.12 (1987), pp. 36–7.
51 *Izvestiia* 17 October 1987, p. 1.
52 *VAN* no.3 (1984), p. 15.
53 *VAN* no.6 (1978), p. 37; no.1 (1987), p. 77; no.4 (1986), p. 41.
54 *VAN* no.6 (1978), p. 37
55 *VAN* no.8 (1977), p. 45; V.I. Kudashov, E.S. Kukolko, and P.D. Gorodets, *Upravlenie issledovaniiami i razrabotkami v akademicheskom nauchno-tekhnicheskom komplekse*, Nauka i tekhnika, Minsk, 1986, p. 125.
56 *VAN* no.8 (1977), pp. 41–8; Kudashov, *Upravlenie issledovaniiami*, p. 126.
57 H. Balzer, *Soviet Research and Development. Information and insights from the third emigration*, final report to the National Council for Soviet and East European Research, Harvard University, 1 July 1985, p. 69.
58 *VAN* no.7 (1987), p. 62.
59 V.I. Duzhenkov, *Problemy organizatsii nauki (Regional'nye aspekty)*, Nauka, Moscow, 1978, p. 138.
60 Iu.M. Kanygin and V.A. Botvin, *Problemy razvitiia i ispol'zovaniia nauchnogo potentsiala krupnykh gorodov*, Naukova dumka, Kiev, 1980, p. 70.
61 *Sotsialisticheskaia industriia* 10 February 1987, p. 2.
62 This point is also made, with further data, in V.I. Nedil'ko, *Intensifikatsiia nauki: soderzhanie i planovo-finansovye faktory*, Nauka i tekhnika, Minsk, 1984, p. 83.
63 *VAN* no.2 (1988), p. 15.
64 *Voprosy teorii i praktiki*, p. 172. See also *Naukovedenie i informatika* no.24 (1983), p. 13.
65 *Sotsialisticheskaia industriia* 10 February 1987, p. 2.
66 Kosolapov, *Optimizatsiia*, p. 178.
67 *Voprosy teorii i praktiki*, p. 68.
68 *Pravda* 6 February 1969, p. 2.
69 According to a Leningrad survey laboratory and department heads spend 11.6 per cent of the working day in meetings. A Novosibirsk survey showed that researchers attend 15 to 17 meetings per month, taking 30 per cent of their time. *Ekonomicheskaia gazeta* no.51 (1966), p. 6; *Izvestiia* 11 July 1964, p. 3.

70 Spending hours in the anterooms of superiors is a peculiar Soviet, or rather Russian, habit. Even such eminent academicians as Otto Shmidt and Andrei Kolmogorov could spend hours in the anteroom of the Academy of Sciences. V. Dorofeeva and V. Dorofeev, *Vremia, uchenye, sversheniia*, Politizdat, Moscow, 1975, p. 99.
71 For some data, see *Pravda* 6 February 1969, p. 2.
72 *Izvestiia* 7 April 1987, p. 3.
73 *Izvestiia* 21 May 1988, p. 1. For data for individual institutes, see *VAN* no.1 (1987), p. 64; *Izvestiia* 4 May 1987, p. 2.
74 *Izvestiia* 12 May 1988, p. 3; 15 June 1988, p. 2.
75 *Ekonomicheskaia gazeta* no.51 (1966), p. 6; *Pravda* 6 February 1969, p. 2.
76 *Ekonomicheskaia gazeta* no.51 (1966), p. 6. See also *Ekonomicheskaia gazeta* no.24 (1967), p. 7.
77 Fortescue, *The Communist Party*, pp. 46–7; *Sotsialisticheskaia industriia* 8 December 1985, p. 2.
78 *Osnovy upravleniia naukoi. Ekonomicheskie problemy*, Nauka, Moscow, 1983, pp. 149–50.
79 V.A. Pokrovskii, *Uskorenie nauchno-tekhnicheskogo progressa. Organizatsiia i metody*, Ekonomika, Moscow, 1983, p. 85.
80 L.L. Lubrano, 'Survey research as a source of information for Soviet science policy', *Soviet Union/Union Sovietique* vol.9, pt 1 (1982), p. 69; *VAN* no.7 (1982), pp. 112–13.
81 *Ideino-politicheskoe vospitanie nauchno-tekhnicheskoi intelligentsii. Opyt i problemy*, Nauka, Moscow, 1982, p. 81.
82 *Komsomol'skaia pravda* 21 October 1973, p. 2. For a similar letter, see *Sovetskaia Rossiia* 12 January 1986, p. 3.
83 See Julius Roth's description of 'hired hand' research, cited in W. Broad, and N. Wade, *Betrayers of the Truth. Fraud and deceit in the halls of science*, Century, London, 1983, p. 151.
84 *Pravda* 14 February 1987, p. 3; *Literaturnaia gazeta* 21 January 1987; 8 July 1987, p. 12. For the purge of other leading Kazakh science managers, see *Pravda* 20 August 1986, p. 3.
85 *Voprosy ekonomiki* no.3 (1986), p. 56; *VAN* no.11 (1986), p. 28.
86 A May 1985 decree tried to loosen up the career and pay structure further, by increasing the number of grades of research worker from two to five. We still have no information on how pay scales have been affected. *Ekonomicheskaia gazeta* no.29 (1985), p. 3.
87 *VAN* no.11 (1986), pp. 35–7; *Sotsialisticheskaia industriia* 15 May 1987, p. 2; *Ekonomicheskaia gazeta* no.1 (1987), p. 2.
88 *Pravda* 24 May 1987, p. 1. See also *Moscow News* no.30 (1987), p. 12; *Sotsialisticheskaia industriia* 15 May 1987, p. 2.
89 For complaints of *uravnilovka* in the new system, see *Sotsialisticheskaia industriia* 15 May 1987, p. 2.
90 For evidence of rank-and-file scepticism, see *VAN* no.2 (1988), p. 13.
91 For example, *Sotsialisticheskaia industriia* 20 October 1985, p. 3.

Glossary

CC-CofM	Central Committee (of the CPSU)-Council of Ministers (referring to joint decrees of those two bodies)
edinonachalie	one-man management (the principle of Soviet management by which a single individual bears ultimate decision-making power)
EFRNT	*edinyi fond dlia razvitiia nauki i tekhniki*, single fund for the development of science and technology
ENIMS	Experimental Research Institute for Metalcutting Machine Tools (the head institute of the Ministry of Machine Tool Industry)
fondooruzhennost'	a measure of fixed capital per research staff member
FRPNT	*fond dlia razvitiia proizvodstva, nauki i tekhniki*, fund for the development of production, science, and technology
GKNT	State Committee for Science and Technology
goszakaz *gosudarstvennyi zakaz*	state order (an obligatory plan task given to enterprises or institutes)
GPO	*gosudarstvennoe proizvodstvennoe upravlenie*, state production administration (a middle-level administrative unit of ministries)
glavk	*glavnoe upravlenie*, chief administration (a middle-level administrative unit of ministries)
khozraschet	an accounting system in which a balance of costs and receipts is drawn up. Usually it is expected that the latter should cover the former
komsomol	Communist Union of Youth or Young

	Communist League (the youth wing of the Communist Party)
korrektirovka	the practice of regularly adjusting plans during their implementation
melkotem'e	an excessive number of minor projects in an institute's research plan
MNTK	*mezhotraslevoi nauchno–tekhnicheskii kompleks*, inter-branch scientific–technical complex
naukoved	a specialist in science studies
nomenklatura	a list of posts, appointment to which is the responsibility of a specific party organisation
NPO	*nauchno–proizvodstvennoe ob"edinenie*, science–production association (scientific and production organisations combined in a single unit under the leadership of the scientific organisation)
NTO	*Nauchno-tekhnicheskoe Ob"edinenie*, Scientific–technical Association (the instrument-making enterprise of the Academy of Sciences)
PO	*proizvodstvennoe ob"edinenie*, production association (a number of production enterprises combined in a single unit, with scientific organisations also possible members)
R&D	research and development
RDI	research, development, and innovation
S&T	science and technology
sovmestitel'stvo	simultaneously holding a research and a teaching post
STC	Scientific–Technical Council (of a ministry)
TA	Technical Administration (of a ministry)
VPO	*vsesoiuznoe promyshlennoe ob"edinenie*, all-Union industrial association (a middle-level administrative unit of ministries)
val	gross output
vedomstvennost'	departmentalism or sectionalism (the illegitimate protection of their own interests by administrative agencies)
vnedrenie	the introduction of the results of research into production
vuz	*vysshoe uchebnoe zavedenie*, higher educational establishment

zakazchik	**orderer (the organisation which orders a piece of research)**
zakaz-nariad	**work order (assignment in the form of a contract between a ministry and a subordinate R&D organisation)**

References

Afanas'ev, V.G. *Nauchno-tekhnicheskaia revoliutsiia, upravlenie, obrazovanie*, Politizdat, Moscow, 1972.

Agursky, M. *The Research Institute of Machine-Building Technology. A part of the Soviet military-industrial complex*, Soviet Institutions Series Paper No.8, The Hebrew University of Jerusalem, September 1976.

Ailes, C.P. and Pardee, A.E. *Cooperation in Science and Technology: an evaluation of the US–Soviet agreement*, Westview, Boulder, CO, 1986.

——and Rushing, F.W. *The Science Race. Training and utilization of scientists and engineers, US and USSR*, Crane Russak, New York, 1982.

Aivazian, S.I., Vedeneev, Iu.A., and Supataeva, O.A. *Pravovye voprosy upravleniia nauchnymi issledovaniiami*, Nauka, Moscow, 1980.

Akademik I.I. Artobolevskii, Znanie, Moscow, 1983.

Alexander, A.J. 'Soviet weapons development and the scientific community', in *Science and Technology in the Soviet Union. Proceedings of a conference*, Department of Statistics, Stanford University, January 1985, chapter 2.

Amann, R. and Cooper, J. (eds) *Industrial Innovation in the Soviet Union*, Yale University Press, New Haven, CN, and London, 1982.

——(eds) *Technical Progress and Soviet Economic Development*, Blackwell, Oxford, 1986.

——and Davies, R.W. (eds) *The Technological Level of Soviet Industry*, Yale University Press, New Haven, CN, and London, 1977.

Andreev, I.D. *Puti povysheniia effektivnosti nauchnogo truda*, Nauka, Moscow, 1985.

Andronikashvili, E. *Nachinaiu s El'brusa, Tvorcheskie portrety uchenykh*, Metsnieraba, Tbilisi, 1982.

Arkhangel'skii, V.N. *Territorial'noe upravlenie razvitiem nauki i tekhniki*, Ekonomika, Moscow, 1985.

Aver'ianov, B.V. *Funktsii i organizatsionnaia struktura organa gosudarstvennogo upravleniia*, Naukova dumka, Kiev, 1979.

Azbel, M.Ya. *Refusenik. Trapped in the Soviet Union*, Hamish Hamilton, London, 1982.

Bakunin, A.V. (ed.) *Deiatel'nost' KPSS po uskoreniiu nauchno-tekhnicheskogo progressa*, Vysshaia shkola, Moscow, 1980.

Balzer, H. *Soviet Research and Development. Information and insights from the third emigration*, final report to the National Council for Soviet and East European Research, Harvard University, 1 July 1986.

Beliaev, E.A. *KPSS i organizatsiia nauki SSSR*, Politizdat, Moscow, 1982.

Belitsky, E. *The Central Institute of Mathematical Economics*, Soviet Institutions Series, Paper No.1, The Hebrew University of Jerusalem, 1974.

Belkin, P.G., Emel'ianov, E.N., and Ivanov, M.A. *Sotsial'naia psikhologiia nauchnogo kollektiva*, Nauka, Moscow, 1987.

Berg, R. *Sukhovei. Vospominaniia genetika*, Chalidze, New York, 1983.

Berliner, Iu.I. *Vnedrenie nauchno-tekhnicheskikh razrabotok. Problemy planirovaniia*, Ekonomika, Moscow, 1985.

Berry, M.J. (ed.) *Science and Technology in the USSR*, Longman, Harlow, 1988.

Bliakhman, L.S. and Mintairov, M.S. (eds) *Normativnye osnovy planirovaniia nauchno-tekhnicheskogo progressa*, Nauka, Leningrad, 1981.

Brezhnev, L.I. *Tselina*, Politizdat, Moscow, 1978.

Broad, W. and Wade, N. *Betrayers of the Truth. Fraud and deceit in the halls of science*, Century, London, 1983.

Campbell, R.W. *The Economics of Oil and Gas*, Johns Hopkins University Press, Baltimore, MD, 1968.

——'Management spillovers from Soviet space and military programmes', *Soviet Studies* vol.23, no.4, April 1972, pp.586–607.

'The case of Igor Melchuk', *Survey* vol.23, no.2 (103) (Spring 1977–8), pp.126–40.

CIA Scientific Institutions. *USSR State Committee for Science and Technology. A reference aid*, CR81-11110, Central Intelligence Agency, Washington, DC, June 1981.

Cocks, P.M. *Science Policy: USA/USSR*, vol.2, *Science policy in the USSR*, National Science Foundation, Washington, DC, 1980.

Cooper, J. 'Civilian and military engineering', in Berry, M.J. (ed.) *Science and Technology in the USSR*, Longman, Harlow, 1988, chapter 22.

—— 'The civilian production of the Soviet defence industry', in Amann, R. and Cooper, J. (eds) *Technical Progress and Soviet Economic Development*, Blackwell, Oxford, 1986.

—— *The Scales of Output of Civilian Products by Enterprises of the Soviet Defence Industry*, CREES Discussion Paper, Soviet Industry and Technology Series No.3, Centre for Russian and East European Studies, University of Birmingham, August 1988.

——'The Soviet Union', in McLean, S. (ed.) *How Nuclear Decisions are Made*, Macmillan, London and Basingstoke, 1986, chapter 1.

——'Western technology and the Soviet defense industry', in Parrott, B. (ed.) *Trade, Technology, and Soviet–American Relations*, Indiana University Press, Bloomington, IN, 1985, pp.169–202.

Davitnidze, I.L. *Kollegii ministerstv (Pravovoe polozhenie i organizatsiia raboty)*, Iuridizdat, Moscow, 1972.

Dorofeeva, V. and Dorofeev, V. *Vremia, uchenye, sversheniia*, Politizdat, Moscow, 1975.

Dubrovskii, K. *Otsenka i upravlenie nauchnymi issledovaniiami v otrasli promyshlennosti*, presentation to an international symposium of scientists and specialists of CMEA countries and Yugoslavia, 'NTP i sotsial'nyi progress', section 5, 'Problemy otsenki i effektivnosti nauchno-tekhnicheskikh razrabotok i novoi tekhniki', Moscow, 1974.

Duzhenkov, V.I. *Problemy organizatsii nauki (Regional'nye aspekty)*, Nauka, Moscow, 1978.

Ekonomicheskoe upravlenie naukoi, mezhotraslevye voprosy, proceedings of symposium 'Sotsial'nye i ekonomicheskie aspekty povysheniia effektivnosti sovetskoi nauki', Zvenigorod, 24–6 May 1982.

Fortescue, S. *The Academy Reorganized. The R&D role of the Soviet Academy of Sciences since 1961*, Occasional Paper No.17, Department of Political Science, Research School of Social Sciences, Australian National University, Canberra, 1983.

——*The Communist Party and Soviet Science*, Macmillan, London and Basingstoke, 1986.

——'Party membership in Soviet research institutes', *Soviet Union/Union Sovietique* vol.11, pt 2 (1984), pp.129–56

——'Party secretaries in Soviet research institutes: divided loyalties?', *Politics*, Australia., vol.18, no.1 (May 1983), pp.73–83.

——'Project planning in Soviet R&D', *Research Policy* vol.14 (1985), pp.267–82.

——'Recent developments in the administration of branch military research', in Sinclair, C. (ed.) *The Status of Soviet Civil Science*, Nijhoff, Dordrecht, 1987, chapter 2.3.

——'The regional party apparatus in the "sectional society"', *Studies in Comparative Communism* vol.21, no.1 (Spring 1988), pp.11–23.

——'Research party organisations and the right of control', *Soviet Studies* vol.35, no.2 (April 1983), pp.175–95.

——'The Soviet Academy of Sciences under Gorbachev', *Australian Slavonic and East European Studies* vol.1, no.2 (1987), pp.1–28.

Frame, J.D. and Baum, J.J. 'Cross-national information flows: examples taken from physics', *Journal of the American Society for Information Science* vol.29 (September 1978), pp.247–52.

——and Carpenter, M.P. 'International research collaboration', *Social Studies of Science* vol.9, no.4 (December 1979), pp.481–97.

——and Narin, F. 'The international distribution of biomedical publications', *Federation Proceedings* vol.36 (May 1977), pp.1790-5.

Freidzon, S. *Top-Level Administration of the Soviet Economy: a partial view*, Rand Paper P-7178, Santa Monica, CA, January 1986.

Frenkel', B.Ia. *Iakov Il'ich Frenkel'*, Nauka, Moscow–Leningrad, 1966.

Garwin, R.L. *A Review of Cooperation in Science and Technology between the US and USSR*, National Academy of Sciences, Washington, DC, 1977.

Glazov, Iu. *Tesnye vrata. Vozrozhdenie russkoi intelligentsii*, Overseas Publications Interchange, London, 1985.

Golovin, I. *Igor' Kurchatov*, Mir, Moscow, 1969.

Goodman, S.E. and McHenry, W.K. 'Computing in the Soviet general economy: an introductory overview', in *Science and Technology in the Soviet Union. Proceedings of a conference*, Department of Statistics,

Stanford University, January 1985.
Gordeeva, N.A. *Vzaimosviazi akademicheskoi nauki i proizvodstva. Organizatsionno-pravovoi aspekt*, Nauka, Moscow, 1985.
Gosudarstvennoe upravlenie nauchno-tekhnicheskim progressom v promyshlennosti soiuznoi respubliki (na materialakh Kazakhskoi SSR), Nauka, Alma-Ata, 1984.
Graham, L.R. 'The role of the Academy of Sciences', *Survey* vol.23, no.1 (102) (Winter 1977–8), pp.117–33.
——*Science, Philosophy, and Human Behavior in the Soviet Union*, Columbia University Press, New York, 1987.
——*The Soviet Academy of Sciences and the Communist Party, 1927–1932*, Princeton University Press, Princeton, NJ, 1967.
Greenberg, L.L. 'Policy-making in the USSR Academy of Sciences' *Journal of Contemporary History*, vol.8, no.4 (October 1973), pp.67–80.
Grinberg, A.P. and Frenkel', B.Ia. *Igor' Vasil'evich Kurchatov v Fiziko-Tekhnicheskom Institute*, Nauka, Leningrad, 1984.
Gusakov, M.A. (ed.) *Organizatsionnye formy sviazi nauki s proizvodstvom v promyshlennosti*, Nauka, Leningrad, 1980.
Gustafson, T. 'Why doesn't Soviet science do better than it does?', in Lubrano, L.L. and Solomon, S.G. (eds) *The Social Context of Soviet Science*, Westview, Boulder, CO, and Dawson, Folkestone, 1980.
Gvishiani, D.M. (ed.) *Upravlenie, planirovanie i organizatsiia nauchnykh i tekhnicheskikh issledovanii*, VINITI, Moscow, 1970, vol.4.
Hanson, P. *Soviet Industrial Espionage: some new information*, CREES, University of Birmingham, 1986.
——and Pavitt, K. *The Comparative Economics of Research, Development and Innovation in East and West: a survey*, Science Policy Research Unit, University of Sussex, Brighton, February 1987.
Holloway, D. 'Innovation in the defence sector', in Amann, R. and Cooper, J. (eds) *Industrial Innovation in the Soviet Union*, Yale University Press, New Haven, CN, and London, 1982, chapter 7.
——*The Soviet Union and the Arms Race*, Yale University Press, New Haven, CN, and London, 1983.
Hough, J.F. 'The historical legacy in Soviet weapons development', in Valenta, J. and Potter, W.C. (eds) *Soviet Decision-making for National Security*, Allen and Unwin, London, 1984.
——*The Soviet Prefects. The local party organs in industrial decision-making*, Harvard University Press, Cambridge, MA, 1969.
Ideino-politicheskoe vospitanie nauchno-tekhnicheskoi intelligentsii. Opyt i problemy, Nauka, Moscow, 1982.
I.P. Bardin v vospominaniiakh sovremennikov, Nauka, Moscow, 1985.
Irvine, J. and Martin, B.R. 'Basic research in the East and West: a comparison of the scientific performance of high-energy accelerators', *Social Studies of Science* vol.15, no.2 (May 1985), pp.293–341.
Kadry – reshaiushchaia sila partiinogo rukovodstva, Mysl', Moscow, 1984.
Kalimanov, A.N. *Organizatsiia i planirovanie nauchno-tekhnologicheskoi deiatel'nosti v tselliulozno-bumazhnoi promyshlennosti*, Lesnaia promyshlennost', Moscow, 1983.

Kalinin, N.G. *General'nye skhemy upravleniia*, Znanie, Moscow, 1982.
——(ed.) *Organizatsiia upravleniia v sisteme ministerstv*, Izd-vo Moskovskogo un-ta, Moscow, 1974.
Kanygin, Iu.M. and Botvin, V.A. *Problemy razvitiia i ispol'zovaniia nauchnogo potentsiala krupnykh gorodov*, Naukova dumka, Kiev, 1980.
Kaplan, N. 'Research administration and the administrator: USSR and US', *Administrative Science Quarterly* vol.6 (1961), pp.51–72.
Kaysen, C. *Review of US–USSR Interacademy Exchanges and Relations*, National Academy of Sciences, Washington, DC, 1977.
Kelly, W.J., Shaffer, H.L., and Thompson, J.K. *Energy Research and Development in the USSR*, Duke University Press, Durham, NC, 1986.
Khrushchev Remembers. The last testament, Deutsch, London, 1974.
Komkov, G.D., Levshin, B.V., and Semenov, L.K. *Akademiia nauk SSSR. Kratkii istoricheskii ocherk*, 2nd edn, 2 vols, Nauka, Moscow, 1977.
Kosolapov, V.V. and Shcherban', A.N. *Optimizatsiia nauchno-issledovatel'skoi deiatel'nosti*, Naukova dumka, Kiev, 1971.
Kozlov, B.I. (ed.) *Organizatsiia i razvitie otraslevykh nauchno-issledovatel'skikh institutov Leningrada. 1917–1977*, Nauka, Leningrad, 1979.
Kuchment, M. and Balzer, H. *A Net Assessment of Soviet R&D Capabilities: information and insights from the third emigration*, Russian Research Center, Harvard, 1981.
Kudashov, V.I., Kukolko, E.S., and Gorodets, P.D. *Upravlenie issledovaniiami i razrabotkami v akademicheskom nauchno-tekhnicheskom komplekse*, Nauka i tekhnika, Minsk, 1986.
Kukrus, A. *Nauka i upravlenie (upravlenie naukoi v SSSR)*, Valgus, Tallin, 1982.
Lakhtin, G.A. *Upravlenie v nauchnom uchrezhdenii*, Energoatomizdat, Moscow, 1983.
Lesbourne, J. and Sonnenschein, H. (eds) *Fundamentals of Pure and Applied Economics*, Harvard Academic Publishers, forthcoming.
Levshin, L.V. *Aleksandr Nikolaevich Terenin 1896–1967*, Nauka, Moscow, 1985.
Lewis, R.A. 'Hierarchy and technological innovation in Soviet industry: the science–production associations', *Minerva* vol.22, no.2 (Summer 1984), pp.129–59.
——*The Research and Development Effort in the Soviet Union. The manpower data, 1962–1969*, CREES Discussion Paper, University of Birmingham, October 1971.
Lowenhardt, J. '"The tale of the torch". Scientist entrepreneurs in the Soviet Union', *Survey* vol.20, no.4 (93) (Autumn 1974), pp.113–21.
L.L. Lubrano, 'National and international politics in US–USSR cooperation', *Social Studies of Science* vol.11, no.4 (November 1981), pp.451–80.
——'Soviet science specialists: professional roles and policy involvement', in Remnek, R.B. (ed.) *Social Scientists and Policy Making in the USSR*, Praeger, New York, 1977.
——'Survey research as a source of information for Soviet science policy', *Soviet Union/Union Sovietique* vol.9, pt 1 (1982), pp.55–81.

——and Solomon, S.G. (eds) *The Social Context of Soviet Science*, Westview, Boulder, CO, and Dawson, Folkestone, 1980.

McLean, S. (ed.) *How Nuclear Decisions are Made*, Macmillan, London and Basingstoke, 1986.

Marsh, R.J. *Soviet Fiction since Stalin. Science, politics and literature*, Croom Helm, London and Sydney, 1986.

Medvedev, Z.A. *The Medvedev Papers*, Macmillan, London, 1971.

——*The Rise and Fall of T.D. Lysenko*, Columbia University Press, New York and London, 1969.

——*Soviet Science*, Oxford University Press, Oxford, 1979.

Metodologicheskie problemy nauchno-tekhnicheskogo progressa. Voprosy integratsii nauki i proizvodstva, Nauka, Novosibirsk, 1987.

Metodologicheskie problemy razvitiia ekonomicheskikh sviazei v sisteme 'Nauka–proizvodstvo', Moscow, 1983.

Mikulinskii, S.R. and Rikhta, R. (eds) *Sotsializm i nauka*, Nauka, Moscow, 1981.

Miller, R.F. 'The role of the Communist Party in Soviet research and development', *Soviet Studies* vol.37, no.1 (January 1985), pp.31–59.

Mil'ner, B.Z. (ed.) *Sovershenstvovanie upravleniia proizvodstvom novoi tekhniki*, Nauka, Moscow, 1984.

Narin, F. and Carpenter, M.P. 'National publication and citation comparisons', *Journal of the American Society for Information Science* vol.26 (1975), pp.80–93.

——Frame, J.D., and Carpenter, M.P. 'Highly cited Soviet papers: an exploratory investigation', *Social Studies of Science* vol.13 (1983), pp.307–19.

Narodnoe khoziaistvo SSSR za 70 let. 1917–1987. Iubileinyi statisticheskii ezhegodnik, Finansy i statistika, Moscow, 1987.

Nauchno-tekhnicheskaia intelligentsiia razvitogo sotsializma, Kemerovskii gos. un-t, Kemerovo, 1985.

Nedil'ko, V.I. *Intensifikatsiia nauki: soderzhanie i planovo-finansovye faktory*, Nauka i tekhnika, Minsk, 1984.

Nolting, L.E. *Financing of Research, Development and Innovation in the USSR*, Foreign Economic Report No.22, Department of Commerce, Washington, DC, September 1985.

——*Management and Financing of Research, Development, and Innovation in the Soviet Electrotechnical Industry*, Center for International Research Staff Paper No.43, US Bureau of the Census, Washington, DC, August 1988.

——*The 1968 Reform of Scientific Research, Development and Innovation in the USSR*, US Department of Commerce, Washington, DC, 1976.

——*The Planning of Research, Development, and Innovation in the USSR*, Foreign Economic Report No.14, US Department of Commerce, Washington, DC, 1978.

——*The Soviet View of the State of Technological Innovation in the USSR*, Center for International Research Staff Paper No.38, US Bureau of the Census, Washington, DC, 1988.

O korennoi perestroike upravleniia ekonomikoi. Sbornik dokumentov,

Politizdat, Moscow, 1987.
Organizatsiia i razvitie otraslevykh nauchno-issledovatel'skikh institutov Leningrada, 1917–1977, Nauka, Leningrad, 1979.
Organizatsiia promyshlennykh issledovanii i razrabotok novoi tekhniki, Lenizdat, Leningrad, 1978.
Osnovy upravleniia naukoi. Ekonomicheskie problemy, Nauka, Moscow, 1983.
Parrott, B. (ed.) *Trade, Technology, and Soviet–American Relations*, Indiana University Press, Bloomington, IN, 1985.
Partiia ukrepliaet svoi riady, Mysl', Moscow, 1985.
Partiinye organizatsii Urala i nauchno-tekhnicheskii progress, Ural'skii gos. un-t, Sverdlovsk, 1985.
Planirovanie, tekhnicheskii progress, effektivnost', Ekonomika, Moscow, 1985.
Pokrovskii, V.A. *Uskorenie nauchno-tekhnicheskogo progressa. Organizatsiia i metody*, Ekonomika, Moscow, 1983.
Popovsky, M. *Manipulated Science*, Doubleday, New York, 1979.
——*Science in Chains. The crisis of science and scientists in the Soviet Union today*, Collins and Harvill, London, 1980.
——*Upravliaemaia nauka*, Overseas Publications Interchange, London, 1978.
Problemy rukovodstva nauchnym kollektivom. Opyt sotsial'no-psikhologicheskogo issledovaniia, Nauka, Moscow, 1982.
Pronina, V.S. *Tsentral'nye organy upravleniia narodnym khoziaistvom*, Iuridizdat, Moscow, 1971.
Rabkin, Yu.M. 'Soviet organization of international scientific contacts', in Sinclair, C. (ed.) *The Status of Soviet Civil Science*, Nijhoff, Dordrecht, 1987.
Rajana, C. *The Chemical and Petro-Chemical Industries of Russia and Eastern Europe, 1960–1980*, Sussex University Press, Brighton, 1975.
Rassokhin, V.P. *Mekhanizm vnedreniia dostizhenii nauki. Politika, upravlenie, pravo*, Nauka, Moscow, 1985.
Remnek, R.B. (ed.) *Social Scientists and Policy Making in the USSR*, Praeger, New York, 1977.
Resheniia partii i pravitel'stva po khoziaistvennym voprosam. Sbornik dokumentov. Tom 11, Noiabr' 1975 g. - iun' 1977 g., Politizdat, Moscow, 1977.
Richardson, J. 'The Soviet ministerial élite in the Brezhnev era', seminar paper, Department of Political Science, Research School of Social Sciences, Australian National University, Canberra, 7 June 1982.
Romanov, A.K., Androsova, L.A., and Felinger, A.F. *Nauchnye kadry Sibirskogo otdeleniia AN SSSR. Metody i rezul'taty statisticheskogo issledovaniia*, Nauka, Novosibirsk, 1979.
Sagdeev, R. 'Perestroika and the scientific intelligentsia', Meeting report, Kennan Institute for Advanced Russian Studies, Washington, DC, 16 November 1988.
Sakharov, A. 'Fragmenty iz avtobiograficheskoi knigi', *Kontinent*, no.36 (1983), pp.51–60.
Science and Technology in the Soviet Union. Proceedings of a conference, Department of Statistics, Stanford University, January 1985.

Science Indicators. The 1985 report, National Science Board, Washington, DC, 1985.
Shcherbakov, A.I. *Effektivnost' nauchnoi deiatel'nosti v SSSR*, Ekonomika, Moscow, 1982.
Shcherbitskii, B.V. and Tarasovich, V.S. (eds) *Planovoe upravlenie razvitiem nauki i tekhniki v soiuznoi respublike*, Naukova dumka, Kiev, 1981.
Sigov, I.I. (ed.) *Kompleksnaia programma nauchno-tekhnicheskogo progressa v regione*, Nauka, Leningrad, 1983.
Sinclair, C. (ed.) *The Status of Soviet Civil Science*, Nijhoff, Dordrecht, 1987.
Sliusarenko, A.I. *KPSS i progress nauki v razvitom sotsialisticheskom obshchestve*, Vishcha shkola, Kiev, 1978.
Sominskii, V.S., Kublis, V.I., and Fedorov, K.G. *Upravlenie naukoi v khimicheskoi promyshlennosti*, Khimiia, Moscow, 1978.
Sotsial'nye i ekonomicheskie problemy povysheniia effektivnosti nauki, Nauka, Moscow, 1985.
Soviet Acquisition of Militarily Significant Western Technology: an update, USGPO, Washington, DC, 1985.
Suttmeier, R.P. 'Socialist science policy', *Problems of Communism* vol.26, no.4 (July–August 1977), pp.72–7.
Sutton, J.R. 'Organizational autonomy and professional norms in science: a case study of the Lawrence Livermore Laboratory', *Social Studies of Science* vol.14, no.2 (May 1984), pp.197–224.
Taksir, K.I. *Nauchno–proizvodstvennye ob"edineniia*, Nauka, Moscow, 1977.
Tedstrom, J. 'The economics of Soviet defense spending', *Radio Liberty Research*, RL Supplement 4/88 (15 August 1988).
Teoriia i praktika organizatsii kompleksnykh issledovanii, Nauka, Leningrad, 1986.
Thomas, J.R. 'Militarization of the Soviet Academy of Sciences', *Survey* vol.29, no.1 (124) (Spring 1985), pp.29–58.
——and Kruse-Vaucienne, U.M. (eds) *Soviet Science and Technology: domestic and foreign perspectives*, George Washington University, 1977.
Trapeznikov, V.A. (ed.) *Upravlenie razvitiem nauki i tekhniki*, Ekonomika, Moscow, 1980.
Uchenye i nauchnyi kollektiv: sotsial'nye aspekty deiatel'nosti, Progress, Moscow, 1986.
Valenta, J. and Potter, W.C. (eds.) *Soviet Decision-making for National Security*, Allen and Unwin, London, 1984.
Veger, L.L. *Ekonomicheskii effekt i upravlenie NIOKR*, Nauka, Moscow, 1985.
Vishniakov, V.G. *Struktura i shtaby organov sovetskogo gosudarstvennogo upravleniia*, Nauka, Moscow, 1972.
Voprosy teorii i praktiki upravleniia i organizatsiia nauki, Nauka, Moscow, 1975.
Vorotenitskaia, S.L. *Upravlenie kachestvom produktsii v pishchevoi promyshlennosti*, Izd-vo standartov, Moscow, 1983.
Voslensky, M. *Nomenklatura. Anatomy of the Soviet ruling class*, Bodley Head, London, 1984.
Vospominaniia ob akademike L.A. Artsimoviche, Nauka, Moscow, 1981.

Vospominaniia o I.E. Tamme, 2nd edn, Nauka, Moscow, 1986.
Vucinich, A. *Empire of Knowledge. The Academy of Sciences of the USSR (1917–1970)*, University of California Press, Berkeley and Los Angeles, CA, 1984.
Wienert, H. 'The organisation and planning of research in the higher educational establishments', in Zaleski, E. et al. (eds) *Science Policy in the USSR*, OECD, Paris, 1969.
Wirszup, I. 'Soviet secondary school mathematics and science programs', in Sinclair, C. (ed.) *The Status of Soviet Civil Science*, Nijhoff, Dordrecht, 1987.
Wolton, T. *Le KGB en France*, Grasset, Paris, 1986.
Woods, S. 'Weapons acquisition in the Soviet Union', *Soviet Union/Union Sovietique* vol.10, pts 2–3 (1983), pp.337–73.
Zaleski, E., Kozlowski, J.P., Wienert, H., Davies, R.W., Berry, M.J., and Amann, R. (eds) *Science Policy in the USSR*, OECD, Paris, 1969.
Zemtsov, I. *IKSI. The Moscow Institute of Applied Social Research. A note on the development of sociology in the USSR*, Soviet Institutions Series, Paper No.6, The Hebrew University of Jerusalem, April 1976.
Zvorykin, A.A. (ed.) *Nauchnyi kollektiv. Opyt sotsiologicheskogo issledovaniia*, Nauka, Moscow, 1980.

Index

For Product Safety Concerns and Information please contact our EU representative GPSR@taylorandfrancis.com
Taylor & Francis Verlag GmbH, Kaufingerstraße 24, 80331 München, Germany

www.ingramcontent.com/pod-product-compliance
Lightning Source LLC
LaVergne TN
LVHW020641300926
849161LV00004B/764
9781032889764